Study Guide and Solutions Ma

CHEMISTRY FO
General, Organic, and Biochemistry

INTRODUCTORY CHEMISTRY FOR TODAY

ORGANIC AND BIOCHEMISTRY FOR TODAY

SECOND EDITION

SPENCER L. SEAGER
Weber State University

MICHAEL R. SLABAUGH
Weber State University

WEST PUBLISHING COMPANY
Minneapolis/St. Paul   New York   Los Angeles   San Francisco

Cover Image © 1993 LeRoy Behling

WEST'S COMMITMENT TO THE ENVIRONMENT

In 1906, West Publishing Company began recycling materials left over from the production of books. This began a tradition of efficient and responsible use of resources. Today, up to 95% of our legal books and 70% of our college texts and school texts are printed on recycled, acid-free stock. West also recycles nearly 22 million pounds of scrap paper annually—the equivalent of 181,717 trees. Since the 1960s, West has devised ways to capture and recycle waste inks, solvents, oils, and vapors created in the printing process. We also recycle plastics of all kinds, wood, glass, corrugated cardboard, and batteries, and have eliminated the use of Styrofoam book packaging. We at West are proud of the longevity and the scope of our commitment to the environment.

Production, Prepress, Printing and Binding by West Publishing Company.

COPYRIGHT © 1995 by   WEST PUBLISHING CO.
　　　　　　　　　　　610 Opperman Drive
　　　　　　　　　　　P.O. Box 64526
　　　　　　　　　　　St. Paul, MN 55164–0526

All rights reserved
Printed in the United States of America
02　01　00　99　98　97　96　95　　8　7　6　5　4　3　2　1

ISBN 0–314–04274–1

CONTENTS

Chapter 1	Matter, Measurements, and Calculations	1
Chapter 2	Atoms and Molecules	17
Chapter 3	Electronic and Nuclear Characteristics	31
Chapter 4	Forces Between Particles	45
Chapter 5	Chemical Reactions	59
Chapter 6	The States of Matter	73
Chapter 7	Solutions and Colloids	93
Chapter 8	Reaction Rates and Equilibrium	115
Chapter 9	Acids, Bases, and Salts	127
Chapter 10	Organic Compounds: Alkanes	151
Chapter 11	Unsaturated Hydrocarbons	167
Chapter 12	Alcohols, Phenols, and Ethers	179
Chapter 13	Aldehydes and Ketones	189
Chapter 14	Carboxylic Acids and Esters	199
Chapter 15	Amines and Amides	211
Chapter 16	Carbohydrates	221
Chapter 17	Lipids	233
Chapter 18	Proteins	243
Chapter 19	Enzymes	253
Chapter 20	Nucleic Acids and Protein Synthesis	263
Chapter 21	Nutrition and Energy for Life	273
Chapter 22	Carbohydrate Metabolism	281
Chapter 23	Lipid and Amino Acid Metabolism	291
Chapter 24	Body Fluids	301

Please note: This manual accompanies the comprehensive volume *Chemistry for Today: General Organic and Biochemistry*, Second Edition and its two split volumes, *Introductory Chemistry for Today*, Second Edition, and *Organic and Biochemistry for Today*, Second Edition. Chapters 1-10 in *Introductory Chemistry for Today*, Second Edition correspond to the same chapter numbers in *Chemistry for Today: General, Organic and Biochemistry*, Second Edition. If you are using *Organic and Biochemistry for Today*, Second Edition, however, please note that chapters 1-15 in that volume correspond to chapters 10-24 in this manual.

TO THE STUDENT

There are many study aids that are helpful to chemistry students. Some aids our students have often requested are (1) an overview of the topics covered in class, (2) completely worked-out solutions to assigned exercises, and (3) samples of typical examination questions.

We have included these aids in this manual. The overview of topics is provided by a programmed review of each chapter in the form of a fill-in-the-blanks activity. The main terms and ideas of each chapter are the answers to the blanks. The correct answers are provided at the end of each chapter. Complete solutions are given for each exercise that is answered in the textbook. These will help you check your approach to problem solving. Lastly, we have provided numerous examples of examination questions. These are taken from examinations we have given to our students during the past several years. These questions illustrate the kinds of things you might be required to know about the various topics. In addition, they illustrate the types of questions that can be used, such as multiple-choice and true-false. Answers to all examination questions are given at the end of each chapter. We hope you will find these aids useful as you pursue your studies of chemistry.

Spencer Seager and Michael Slabaugh

CHAPTER 1

Matter, Measurements and Calculations

PROGRAMMED REVIEW

Section 1.1 What is Matter?

Matter is anything that (a) _____ _____ and (b) _____ _____.
(c) _____ is a measure of the amount of matter present. The gravitational force pulling an object toward the earth is called the object's (d) _____.

Section 1.2 Properties and Changes

(a) _____ properties can be observed or measured without changing or attempting to change the (b) _____ of the matter in question. (c) _____ properties are demonstrated when attempts are made to change the composition of matter. (d) _____ can also be classified as physical or chemical.

Section 1.3 A Model of Matter

Explanations for observed behavior are called scientific (a) _____. A (b) _____ is the smallest unit of a pure substance. Molecules containing one atom are called (c) _____ molecules. Molecules made up of identical atoms are called (d) _____ molecules. The term polyatomic is used to describe molecules containing (e) _____ or more atoms. An atom is the limit of (f) _____ subdivision.

Section 1.4 Classification of Matter

Matter that has a constant composition and fixed properties is called a (a) _____ _____. A (b) _____ is a physical blend of matter that can be physically separated into two or more components. (c) _____ matter has the same appearance and properties throughout. Homogeneous mixtures of two or more substances are called (d) _____. Most matter found in nature is classified as (e) _____. A pure substance is homogeneous, and is either a (f) _____ or an (g) _____.

Section 1.5 Measurement Units

Measurements are based on (a) _____ that have been agreed upon. A measurement is expressed as some (b) _____ of a specific (c) _____. The earliest measurements were based on dimensions of the (d) _____ _____.

Section 1.6 The Metric System

A specific unit from which other units are obtained by multiplication or division is called a (a) _____ _____ of measurement. Units obtained from basic units are called (b) _____ _____. The basic units of length, volume and mass respectively in the metric system are the (c) _____, (d) _____ _____ and (e) _____. Units of the metric system that are larger or smaller than a basic unit are indicated by (f) _____ attached to the basic unit. The basic unit of temperature in the metric system is the (g) _____.

Section 1.7 Large and Small Numbers

In scientific notation, a number is represented as a product of a (a) _____ _____ and (b) _____ raised to a whole number exponent. The standard position for a decimal in scientific notation is to the (c) _____ of the first (d) _____ digit in the nonexponential number. In scientific notation, an exponent of -3 indicates the original decimal position is (e) _____ places to the (f) _____ of the standard position.

Section 1.8 Significant Figures

Every measurement contains an (a) _____ that depends on the measuring device. When a measurement is represented using significant figures correctly, the last number is an (b) _____. The number 0.0219 contains (c) _____ significant figures. When the number 8.42149 is properly rounded to four significant figures, the last number will be (d) _____. When (e) _____ _____ are used in calculations, they do not influence the number of significant figures in the calculated results.

Section 1.9 Using Units in Calculations

The (a) _____-_____ method is a systematic approach to solving numerical problems. The (b) _____ used in the factor-unit method are (c) _____ derived from fixed relationships between quantities. In a factor-unit calculation, the units of the factor cancel out the units of the (d) _____ quantity, and generate the units of the (e) _____ quantity.

Matter, Measurements and Calculations 3

Section 1.10 Percent Calculations

The word percent means per (a) _____ _____. A percent is the (b) _____ of specific items found in a group of (c) _____ such items. A basket of fruit contains 3 oranges, 2 apples and 4 pears. In a calculation of the percent pears in the basket, the "total" in the calculation would be (d) _____.

SOLUTIONS TO EXERCISES ANSWERED IN THE TEXT

1.2   Hitting the ball with a hammer does not lift it, so gravity is not involved. The resistance to motion depends only on the mass which is no different on the earth or the moon.

1.3   Weigh an empty basketball, then fill with air and weigh again.

1.5   a) Since mass does not depend on gravity, your mass would be the same at each location.
      b) Since weight depends on gravity, which decreases with distance from the earth's center, your weight would be less on the mountain top.

1.7   a) Chemical. Changes in taste and color indicate a change in composition.
      c) Physical. The broken glass has the same composition as unbroken glass, it is just in smaller pieces.
      e) Chemical. The color change in the silver indicates a likely composition change.

1.8   a) Physical. The melting point is determined by heating the iron until it melts. When it cools and solidifies, it is still iron; no composition change has taken place.
      c) Chemical. Corrosion involves changes in composition. Corroded (rusted) iron is an example.
      e) Chemical. The fact that the acid has been neutralized implies it is no longer an acid. Thus its composition has been changed.
      g) Physical. Hardness is determined by attempting to scratch or dent one material with another. The material scratched from a larger sample has the same composition as the sample from which it was scratched.
      i) Chemical. Degradation is a word that generally implies changing into simpler substances. This would mean a chemical change would accomplish degradation. Thus, failure to degrade is also a chemical property since an attempt must be made to degrade the material in order to observe the property.

4 CHAPTER 1

1.10 b) The white solid remaining is not succinic acid because a gas was given off by succinic acid to form the solid. Thus, the solid must be a different compound than succinic acid.
   d) Polyatomic because a gaseous material that must contain atoms was given off, and left behind a solid that must also contain atoms.
   Heteroatomic because the gas and solid produced by heating succinic acid are two substances that differ from each other, and must therefore contain some different atoms.

1.12 Polyatomic and heteroatomic. Because only a single product is formed, its molecules must contain at least one atom from each of the two reacting gases.

1.14 Polyatomic and heteroatomic. The molecules of the products, carbon dioxide and water, contain carbon and hydrogen respectively that must have come from the alcohol since oxygen was the only other material reacted.

1.15 b) D is an element because elements are defined as containing homoatomic molecules.
   d) L is a compound because it contains atoms from two different elements. Thus, it is composed of heteroatomic molecules.
   f) X cannot be classified because some treatment other than heating might be required to change it into two or more other substances.

1.16 a) Substance R might be an element, but it is very difficult to be certain, because some other treatment might change it into two or more new substances, which would prove it to be a compound. It is very difficult to absolutely classify a substance as an element by trying to get it to change.
   c) The solid left could be either an element or a compound. Tests would have to be done to see if it could be changed into simpler substances.

1.18 b) Homogeneous. It is composed of a pure substance; an element.
   d) Homogeneous. It is a solution (a homogeneous mixture) of water, sugar, flavorings, etc.
   f) Heterogeneous. It contains water, undissolved dirt, etc. that would settle out with time if the mixture were not agitated.
   h) Homogeneous. Its appearance indicates identical composition, etc. regardless of where the sample is tested.
   j) Heterogeneous. Its appearance shows easily-identified areas with different properties such as color.

1.20 a) Heterogeneous. It is not transparent, a characteristic of solutions.
   c) Homogeneous. Pure saliva is a clear (transparent) liquid.
   e) Homogeneous. Pure perspiration is a clear liquid.

Matter, Measurements and Calculations 5

1.21 The given answer is the solution.

1.23 The given answer is the solution.

1.24 a) Not a metric unit according to Table 1.3
c) A metric unit according to Table 1.3
e) Not a metric unit according to Table 1.3
g) Not a metric unit according to Table 1.3
i) A metric unit according to Table 1.3
k) A metric unit according to Table 1.3

1.26 b) 10,000 m. The prefix kilo means thousand times. Thus, 10 kilometers is (10)(1000) m or 10,000 m.
d) One-millionth of a meter. The prefix micro means one-millionth. Thus, a micrometer means a one-millionth meter.

1.28 $(240 \text{ mL})(\dfrac{1 \text{ L}}{1000 \text{ mL}}) = 0.240 \text{ L}$

$(240 \text{ mL})(\dfrac{1 \text{ cm}^3}{1 \text{ mL}}) = 240 \text{ cm}^3$

1.30 $(1500 \text{ m})(\dfrac{1 \text{ km}}{1000 \text{ m}}) = 1.500 \text{ km}$

1.31 a) Liter is larger. 1 L = 1.057 qt
c) kcal is larger. 1 kcal = 3.97 BTU
e) 65 K. $(65°\text{C})(\dfrac{1 \text{ K}}{1°\text{C}}) = 65 K$

1.32 b) A = (length)(width) = (5.0 m)(2.8 m) = 14 m²

V = (length)(width)(height) = (5.0 m)(2.8 m)(2.1 m) = 29 m³

d) $(1.0 \text{ cm}^3)(\dfrac{1 \text{ mL}}{1 \text{ cm}^3})(\dfrac{1 \text{ dm}^3}{1000 \text{ mL}})(\dfrac{1 \text{ kg}}{1.0 \text{ dm}^3})(\dfrac{1000 \text{ g}}{1 \text{ kg}}) = 1.0 \text{ g}$

Factors came from Table 1.3

f) $(5 \text{ grain})(\dfrac{1 \text{ mg}}{0.015 \text{ grain}}) = 333 \text{ mg}$

Factor came from Table 1.3

6   CHAPTER 1

1.33   Use equation 1.1: $°C = \frac{5}{9}(23°F - 32) = \frac{5}{9}(-9) = -5°C$

Use Equation 1.4: $K = °C + 273 = -5 + 273 = 268K$

1.35 $(4500 \text{ kcal})(\frac{4184 \text{ J}}{1 \text{ kcal}}) = 1.88 \times 10^7 \text{ J}$

Factor came from Table 1.3

$(4500 \text{ kcal})(\frac{3.97 \text{ BTU}}{1 \text{ kcal}}) = 1.79 \times 10^4 \text{ BTU}$

Factor came from Table 1.3

1.36   b) The given answer is the solution
d) The given answer is the solution
f) The given answer is the solution
h) The given answer is the solution

1.37   a) The original decimal position is two places to the right of the standard position. Thus, an exponent of 2 must be used. 3.025×10^2
c) The original decimal position is five places to the right of the standard position. Three of those positions are represented by the "thousand." Thus, an exponent of 5 must be used. 1.25×10^5
e) The original decimal position is six places to the left of the standard position. Thus, an exponent of -6 must be used. 8.13×10^{-6}
g) The "thousand" puts the original decimal position three places to the right of the standard position. Thus, an exponent of 3 must be used. 1.02×10^3
i) The original decimal position is three places to the right of the standard position. Thus, an exponent of 3 must be used. 3.050×10^3

1.39   In .0106 cm, the original decimal position is two places to the left of the standard position. Thus, an exponent of -2 must be used. 1.06×10^{-2} cm. Note the non-significant leading zero is not included.

In .0042 in., the original decimal position is three places to the left of the standard position. Thus, an exponent of -3 must be used. 4.2×10^{-3} in. Note the non-significant leading zeroes are not included.

1.41   The exponent 23 means the original decimal position is 23 places to the right of the standard position. This can only be written without using scientific notation by adding 21 zeroes to the right. 602000000000000000000000

Matter, Measurements and Calculations 7

1.42 In each case, multiply the non-exponential numbers, and round the product to the proper number of significant figures. Add together the exponents on 10 to get the exponent used on the 10 in the final answer.

a) $(3.5 \times 10^2)(1.1 \times 10^3) = (3.5)(1.1) \times (10^2)(10^3) = 3.9 \times 10^{2+3} = 3.9 \times 10^5$

c) $(2.8 \times 10^{-3})(1.9 \times 10^2) = (2.8)(1.9) \times (10^{-3})(10^2) = 5.3 \times 10^{-3+2}$
$= 5.3 \times 10^{-1}$

e) $(6.3 \times 10^{-9})(3.7 \times 10^7) = (6.3)(3.7) \times (10^{-9})(10^7) = 23 \times 10^{-9+7} = 23 \times 10^{-2}$
$= 2.3 \times 10^{-1}$

1.43 a) $(0.0820)(0.116) = (8.20 \times 10^{-2})(1.16 \times 10^{-1}) = (8.20)(1.16) \times (10^{-2})(10^{-1})$
$= 9.51 \times 10^{-2-1} = 9.51 \times 10^{-3}$

c) $(720)(0.010) = (7.20 \times 10^2)(1.0 \times 10^{-2}) = (7.2)(1.0) \times (10^2)(10^{-2})$
$= 7.2 \times 10^{2-2} = 7.2 \times 10^0 = 7.2$

e) $(5280)(12) = (5.280 \times 10^3)(1.2 \times 10^1) = (5.28)(1.2) \times (10^3)(10^1)$
$= 6.3 \times 10^{3+1} = 6.3 \times 10^4$

1.44 In each case divide the non-exponential numbers and round the quotient to the proper number of significant figures. Subtract the exponent of the denominator from the exponent of the numerator to get the exponent used on 10 in the final answer.

b) $\dfrac{8.8 \times 10^3}{4.4 \times 10^{-5}} = \dfrac{8.8}{4.4} \times 10^{3-(-5)} = 2.0 \times 10^{3+5} = 2.0 \times 10^8$

d) $\dfrac{5.1 \times 10^{-2}}{8.6 \times 10^{-7}} = \dfrac{5.1}{8.6} \times 10^{-2-(-7)} = 0.59 \times 10^{-2+7} = 0.59 \times 10^5 = 5.9 \times 10^4$

f) $\dfrac{2.7 \times 10^2}{4.9 \times 10^4} = \dfrac{2.7}{4.9} \times 10^{2-4} = 0.55 \times 10^{-2} = 5.5 \times 10^{-3}$

h) $\dfrac{172}{2.15} = \dfrac{1.72 \times 10^2}{2.15} = \dfrac{1.72}{2.15} \times 10^2 = 0.800 \times 10^2 = 8.00 \times 10^1$

1.45 Use the multiplication and division procedures described in Exercises 1.42 and 1.44.

b) $\dfrac{(7.4 \times 10^{-3})(1.3 \times 10^4)}{5.5 \times 10^{-2}} = \dfrac{(7.4)(1.3)}{5.5} \times 10^{-3+4-(-2)} = 1.7 \times 10^{-3+4+2} = 1.7 \times 10^3$

8   CHAPTER 1

$$d) \frac{9.9 \times 10^8}{(3.7 \times 10^{-2})(8.4 \times 10^{-3})} = \frac{9.9}{(3.7)(8.4)} \times 10^{8-(-2)-(-3)} = 0.32 \times 10^{8+2+3}$$

$$= 0.32 \times 10^{13} = 3.2 \times 10^{12}$$

$$f) \frac{(0.064)(0.38)}{(3.0)(7.6)} = \frac{(6.4 \times 10^{-2})(3.8 \times 10^{-1})}{(3.0)(7.6)} = \frac{(6.4)(3.8)}{(3.0)(7.6)} \times (10^{-2})(10^{-1})$$

$$= 1.1 \times 10^{-2-1} = 1.1 \times 10^{-3}$$

$$h) \frac{0.0049}{(3.4)(0.085)} = \frac{4.9 \times 10^{-3}}{(3.4)(8.5 \times 10^{-2})} = \frac{4.9}{(3.4)(8.5)} \times 10^{-3-(-2)}$$

$$= 0.17 \times 10^{-3+2} = 0.17 \times 10^{-1} = 1.7 \times 10^{-2}$$

1.46   In each case the reading should be estimated to one more decimal than the smallest scale marking.
a) Smallest scale marking = 0.1 cm, estimate to 0.01 cm
c) Smallest scale marking = 1°, estimate to 0.1°
e) Smallest scale marking = 1 mL, estimate to 0.1 mL

1.47   In each case, the estimate is made to one more decimal than the smallest scale marking.
a) Smallest scale division = 1°C, estimate to 0.1°C.
Reading = 29.5°C
c) Smallest scale division = 1 degree, estimate to 0.1 degree.
Reading = 15.5 degrees
e) Smallest scale division = 0.1 mL, estimate to 0.01 mL.
Reading = 0.00 mL

1.48   a) Measured number = 25.42 oz

$$\text{weight of average egg} = \frac{25.42 \text{ oz}}{12 \text{ eggs}} = 2.118 \text{ oz/egg}$$

c) Measured numbers are the numbers of people: 19, 24, 17, 31 and 40
Exact number is the 5 days
Average number of people per day =

$$\frac{19 + 24 + 17 + 31 + 40}{5} = \frac{131}{5} = 26.2 \text{ people/day}$$

1.49   b) 3; trailing zeroes are significant
d) 2; leading zeroes are not significant

Matter, Measurements and Calculations 9

f) 3; leading zeroes are not significant, but zero between 5 and 9 is significant
h) 5; trailing zero is significant
j) 3; trailing zeroes are significant
l) 4; all non-zero numbers are significant

1.50 a) $(3.15)(2.0) = 6.3$; two significant figures to match the two in 2.0

c) $\dfrac{(3.65)(0.9986)}{0.4911} = 7.42$; *three signifcant figures to match the three in 3.65*

e) $\dfrac{(760)(1.0)}{640} = 1.2$; *two significant figures to match the two in 1.0*

g) $\dfrac{(19.3)(100)}{1,000} = 1.93$;

three significant figures to match the three in both 19.3 and 100

i) $\dfrac{(251)(3.1 \times 10^{-1})}{(24)(3.0)} = 1.1$;

two significant figures to match the two in 3.1×10^{-1}, 24 and 3.0.

1.51 b) $3.17 + 0.0012 + 0.0009 = 3.17$; express to second position to right of decimal to match 3.17
d) $(3.21 \times 10^{-2}) + (7.4 \times 10^{-1}) = .0321 + 0.74 = 0.77$; express to second position to right of decimal to match 0.74.
f) $36.14 - 0.1085 = 36.03$; express to second position to right of decimal to match 36.14.

1.52 b) $\dfrac{(0.0019 + 0.0029)(8.241)}{37.8209} = \dfrac{(0.0048)(8.241)}{37.8209} = 1.0 \times 10^{-3}$;

two significant figures used to match the two in 0.0048 which resulted from following addition rules.

d) $\dfrac{35.42 - 30.61}{9.97 - 0.24} = \dfrac{4.81}{9.73} = 0.494$;

three significant figures used to match the three in 4.81 and 9.73.

f) $\dfrac{22.72 - 21.95}{0.314} = \dfrac{0.77}{0.314} = 2.5$;

two significant figures used to match the two in 0.77.

1.54 b) (125,000 BTU)(factor) =   kcal; factor must have BTU units in denominator, and kcal units in numerator.

$$\frac{1\ kcal}{3.97\ BTU}$$

d) (200 cm²)(factor) =   in²; factor must have cm² units in denominator, and in² units in numerator.

$$\frac{(0.394\ in)^2}{(1\ cm)^2} = \frac{0.155\ in^2}{1\ cm^2}$$

f) (50 lb)(factor) =   kg; factor must have lb units in denominator, and kg units in numerator.

$$\frac{1\ kg}{2.20\ lb}$$

h) (2 qt)(factor) =   L; factor must have qt units in denominator, and L units in numerator.

$$\frac{1\ L}{1.057\ qt}$$

1.56 (26 mi)(factor) =   km; factor must have mi units in denominator to cancel the mi in 26 mi, and km units in numerator to generate the desired units of the answer.

$$(26\ \cancel{mi})(\frac{1\ km}{0.621\ \cancel{mi}}) = 42\ km$$

1.58 (600 mL)(factor) =   cups; factor must have appropriate units to cancel mL and generate cups.

$$(600\ \cancel{mL})(\frac{0.0338\ \cancel{oz}}{1\ \cancel{mL}})(\frac{1\ cup}{8\ \cancel{oz}}) = 2.54\ cups$$

1.60 Convert 17.97 kg to lbs.

(17.97 kg)(factor) =   lb; factor must have kg units in denominator and lb units in numerator.

$$(17.97\ \cancel{kg})(\frac{2.20\ lb}{1\ \cancel{kg}}) = 39.5\ lb$$

Your luggage weighs less than the 40 lb limit.

1.62 $(1.39\ \frac{mg}{dL})(factors) = \frac{g}{L}$;

factors must cancel $\frac{mg}{dL}$ *and generate* $\frac{g}{L}$.

$$(1.39\ \frac{mg}{dL})(\frac{1\ g}{1000\ mg})(\frac{10\ dL}{1\ L}) = 0.0139\ \frac{g}{L}$$

1.64 $\quad \% = \dfrac{Part}{Total} \times 100 = \dfrac{\$25}{\$185} \times 100 = 13.5\%$

1.66 $\quad \% = \dfrac{Part}{Total} \times 100 = \dfrac{1.2\ mg}{1.4\ mg} \times 100 = 86\%$

1.68 $\quad Total\ mg = 987.1 + 213.3 + 99.7 + 14.4 + 0.1 = 1314.6\ mg$

$\% = \dfrac{Part}{Total} \times 100$

$\% IgC = \dfrac{987.1\ mg}{1314.6\ mg} \times 100 = 75.09\%$

$\% IgA = \dfrac{213.3\ mg}{1314.6\ mg} \times 100 = 16.23\%$

$\% IgM = \dfrac{99.7\ mg}{1314.6\ mg} \times 100 = 7.58\%$

$\% IgD = \dfrac{14.4\ mg}{1314.6\ mg} \times 100 = 1.10\%$

$\% IgE = \dfrac{0.1\ mg}{1314.6\ mg} \times 100 = 0.008\%$

SELF-TEST QUESTIONS

Multiple Choice

1. Which of the following involves a chemical change?
   a) stretching a rubber band   c) lighting a candle
   b) breaking a stick   d) melting an ice cube

12 CHAPTER 1

2. Which of the following terms could not be properly used in the description of a compound?
   a) solution
   b) homogeneous
   c) pure substance
   d) heteroatomic

3. A solid substance is subjected to a number of tests and observations. Which of the following would be classified as a chemical property of the substance?
   a) it is gray in color
   b) it has a density of 2.04 grams per milliliter
   c) it dissolves in acid and a gas is liberated
   d) it is not attracted to either pole of a magnet

4. Which of the following is an example of heterogeneous matter?
   a) water containing crushed ice
   b) a sample of pure table salt
   c) a sample of salt water
   d) a pure sample of iron

5. When a substance undergoes a physical change which of the following is always true?
   a) it melts
   b) a new substance is produced
   c) heat is evolved
   d) the molecules of the substance remain unchanged

6. Which of the following is *not* a chemical change?
   a) burning of magnesium
   b) exploding of some nitroglycerine
   c) pulverizing of some sulfur
   d) rusting of iron

7. Which of the following is the basic unit of length in the metric system?
   a) centimeter
   b) meter
   c) millimeter
   d) kilometer

8. Which of the following is a derived unit?
   a) calorie
   b) cubic decimeter
   c) joule
   d) kilogram

9. In the number 3.91×10^{-3}, the original decimal position is located
   a) 3 places to the right of standard
   b) 2 places to the right of standard
   c) 3 places to the left of standard
   d) 2 places to the left of standard

10. How many significant figures are included in the number 0.02102?
    a) two
    b) three
    c) four
    d) five

11. 21 students in a class of 116 got a B grade on an exam. What percent of the students in the class got B's?
    a) 21.0
    b) 22.1
    c) 15.3
    d) 18.1

12. What single factor derived from Table 1.3 would allow you to calculate the number of quarts in a 2.0 L bottle of soft drink?
    a) 1.057 quart/1 L
    b) 1 L/1.057 quart
    c) 0.0338 fl.oz./1 mL
    d) 1 mL/0.0338 fl.oz.

13. On a hot day a Fahrenheit thermometer reads 97.3°F. What would this reading be on a Celsius thermometer?
    a) 118°C
    b) 22.1°C
    c) 36.3°C
    d) 143°C

Matching

Match the molecules represented below to the correct classification given on the right.

14.

15.

16.

17.

18.

a) homoatomic and polyatomic
b) homoatomic and monoatomic
c) heteroatomic and polyatomic
d) heteroatomic and monoatomic

Match the type of measurement given as responses to the measurement units given below.

19. kelvin

20. milliliter

21. gram

22. millimeter

23. cubic decimeter

24. kilometer

a) mass
b) volume
c) length
d) temperature

True-False

25. The mass of an object is the same as its weight.

26. A physical property can be observed without attempting any composition changes.

27. The cooking of food involves chemical changes.

28. The smallest piece of water that has the properties of water is called an atom.

29. Carbon monoxide molecules are diatomic and heteroatomic. Thus, they contain two identical atoms.

30. The prefix *milli-* means one thousand times.

31. A pure substance containing sulfur and oxygen atoms must be classified as a compound.

32. One meter is shorter than one yard.

33. The calorie and joule are both units of energy.

34. In scientific notation, the exponent on 10 cannot be larger than 10.

35. The correctly rounded sum resulting from adding 13.0, 1.094, 0.132 will contain five significant figures.

ANSWERS TO PROGRAMMED REVIEW

1.1 a) has mass   b) occupies space   c) mass   d) weight

1.2 a) physical   b) composition   c) chemical   d) changes

1.3 a) models   b) molecule   c) monoatomic   d) homoatomic
     e) two   f) chemical

1.4 a) pure substance b) mixture   c) homogeneous   d) solutions
     e) heterogeneous   f) compound   g) element

1.5 a) units   b) multiple   c) unit   d) human body

1.6 a) basic unit   b) derived units   c) meter   d) cubic decimeter
     e) kilogram   f) prefixes   g) kelvin

1.7 a) nonexponential number   b) 10   c) right   d) nonzero
     e) three   f) left

1.8 a) uncertainty   b) estimate   c) three   d) 1
     e) exact numbers

1.9 a) factor-unit   b) factors   c) fractions   d) known
     e) unknown

1.10 a) one hundred   b) number   c) 100   d) 9

ANSWERS TO SELF-TEST QUESTIONS

1.	c	13.	c	25.	F
2.	a	14.	a	26.	T
3.	c	15.	c	27.	T
4.	a	16.	c	28.	F
5.	d	17.	b	29.	F
6.	c	18.	c	30.	F
7.	b	19.	d	31.	T
8.	a	20.	b	32.	F
9.	c	21.	a	33.	T
10.	c	22.	c	34.	F
11.	d	23.	b	35.	F
12.	a	24.	c		

CHAPTER 2

Atoms and Molecules

PROGRAMMED REVIEW

Section 2.1 Symbols and Formulas

Each element has been assigned a unique (a) _____ and (b) _____. The (c) _____ assigned to elements are based on the element names, and consist of a (d) _____ _____ or a (e) _____ _____ followed by a (f) _____ _____. In the (g) _____ for a compound, each atom in the compound molecule is represented by an elemental (h) _____.

Section 2.2 Masses of Atoms and Molecules

Assigned numbers that indicate how one mass compares to another are called (a) _____ _____. Today, average relative masses of atoms and molecules are called (b) _____ _____ and (c) _____ _____, and are expressed in (d) _____ _____ _____.

Section 2.3 Inside the Atom

Most of the mass of an atom is found in the (a) _____ which is made up of particles called (b) _____ and (c) _____. Of the three fundamental particles, (d) _____ have no charge and a mass of (e) _____ amu, (f) _____ have a +1 charge and a mass of (g) _____ amu, and (h) _____ have a -1 charge and a mass of (i) _____ amu.

Section 2.4 Isotopes

The number of (a) _____ in the nucleus of an atom is given by the atomic number. The (b) _____ _____ of an atom is the sum of the number of protons and neutrons in the nucleus. Atoms having the same atomic number but different mass numbers are called (c) _____.

18 CHAPTER 2

Section 2.5 Isotopes and Atomic Weights

The atomic weight of an element consisting of a single isotope is very nearly the same as the (a) _____ _____ of the isotope. The atomic weight of an element consisting of a mixture of isotopes is the (b) _____ relative mass of the isotopic mixture.

Section 2.6 Counting by Weighing

A sample of 20.18 g of neon (Ne) contains the same number of atoms as a sample of bromine (Br) weighing (a) _____ g.

Section 2.7 The Mole

The number of atoms in a sample of an element that weighs the same in grams as the atomic weight of the element is called a (a) _____, and is equal to (b) _____. One mole of krypton atoms (Kr) weighs (c) _____ grams.

Section 2.8 The Mole and Chemical Formulas

The number of moles of nitrogen atoms (N) in 1.5 moles of N_2O is (a) _____. One mole of N_2O weighs (b) _____ grams and contains (c) _____ grams of nitrogen and (d) _____ grams of oxygen.

SOLUTIONS TO EXERCISES ANSWERED IN THE TEXT

2.1   b) The two atoms in a diatomic molecule of a compound will not be identical.

       d) The molecule should be represented by four identical circles and one that is different from the four.

       f) The molecule should be represented by two circles of one type, six circles of another type, and one circle of a third type.

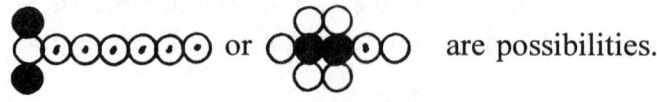 are possibilities.

2.2   a) A diatomic molecule has two atoms. F_2

Atoms and Molecules   19

   c) A triatomic molecule has three atoms. O_3
   e) The number of each type of atom is indicated by a subscript. H_2SO_4

2.3 In each formula, the subscripts indicate the number of atoms in the molecule. The subscript 1 is not written in the formula, but is understood.
   b) The second letter in an elemental symbol is always lower case, so the capital I in silicon should be lower case. $SiCl_4$
   d) The number of atoms in a molecule is represented by a subscript. H_2O_2
   f) The second letter in an elemental symbol is always lower case, so the capital L of chlorine should be lower case. $HClO_2$

2.5 Because the elements combine in a 1:1 atom ratio, the combining masses represent the relative masses of the atoms. To compare them, divide both combining masses by the lowest of the pair.

   Combining mass of K = 3.91
   Combining mass of F = 5.81 - 3.91 = 1.90

   $$\text{Relative mass of } F = \frac{1.90 \text{ g}}{1.90 \text{ g}} = 1.00$$

   $$\text{Relative mass of } K = \frac{3.91 \text{ g}}{1.90 \text{ g}} = 2.06$$

   Thus, on a relative basis, potassium atoms have slightly more than twice the mass of fluorine atoms.

2.7 Because the elements combine in a 1:1 atom ratio, the combining masses represent the relative masses of the atoms. To compare them, divide both combining masses by the lowest of the pair.

   Combining mass of Ca = 2.86 g
   Combining mass of Se = 8.49 - 2.86 = 5.63 g

   $$\text{Relative mass of } Ca = \frac{2.86 \text{ g}}{2.86 \text{ g}} = 1.00$$

   $$\text{Relative mass of } Se = \frac{5.63 \text{ g}}{2.86 \text{ g}} = 1.97$$

   Thus, the ratio of Ca/Se = 1.00/1.97

20   CHAPTER 2

$$\text{Relative mass of Ca} = \frac{40.08}{40.08} = 1.00$$

$$\text{Relative mass of Se} = \frac{78.96}{40.08} = 1.97$$

Thus, the ratio of Ca/Se = 1.00/1.97
The same result as that given by the combining masses data.

2.8 a) The atomic weight of silicon gives the mass of an average atom. An element with an atomic weight of half the atomic weight of silicon would have average atoms of half the mass of silicon atoms. This element is nitrogen, N, with an atomic weight of 14.
c) The atomic weights represent masses of average atoms, so we need two elements with atomic weights that are within 0.3 amu of each other. The two are cobalt (Co) with an atomic weight of 58.93, and nickel (Ni) with an atomic weight of 58.71.
e) The atomic weight of silver as 107.9, so the average silver atom has a mass of 107.9 amu. The atoms we need will have a mass that is 81.2% of the mass of an average silver atom or 87.6 amu. Thus, we need an element with an atomic weight of 87.6. Strontium, Sr, is the needed element.

2.9 Molecular weights are obtained by adding together the atomic weights of all the atoms in a molecule.

   a) O_2 2 x 16.00 amu   =   32.00 amu

   c) $HClO_3$ 1 x 1.008 amu   =   1.008 amu
                         1 x 35.45 amu   =   35.45 amu
                         3 x 16.00 amu   =   48.00 amu
                         Total   =   84.458 amu
                         Rounded Total   =   84.46 amu or 84.5 amu

   e) SO_2 1 x 32.06 amu   =   32.06 amu
                         2 x 16.00 amu   =   32.00 amu
                         Total   =   64.06 amu
                         Rounded Total   =   64.1 amu

Atoms and Molecules

g) CN_2H_4O

1 x 12.01 amu	=	12.01 amu
2 x 14.01 amu	=	28.02 amu
4 x 1.008 amu	=	4.032 amu
1 x 16.00 amu	=	16.00 amu
Total	=	60.062 amu
Rounded Total	=	60.06 amu or 60.1 amu

i) $C_{11}H_{22}O_{11}$

11 x 12.01 amu	=	132.11 amu
22 x 1.008 amu	=	22.176 amu
11 x 16.00 amu	=	176.00 amu
Total	=	330.286 amu
Rounded total	=	330.29 amu or 330.3 amu

2.11 In each case the positive charge of the nucleus will be equal to the number of protons in the nucleus because the neutrons carry zero charge. The mass in amu of each nucleus will equal the total of the number of protons plus the number of neutrons because each particle has a mass of 1 amu.
   a) charge = +2 (from two protons)
      mass = 4 amu (from two protons and two neutrons)
   c) charge = +5 (from five protons)
      mass = 9 amu (from five protons and four neutrons)
   e) charge = +50 (from fifty protons)
      mass = 119 amu (from fifty protons and sixty-nine neutrons)

2.12 In each case the atomic number is equal to the number of protons in the nucleus, and the mass number is equal to the total of the number of protons and neutrons.
   a) atomic number = 2 (from two protons)
      mass number = 4 (from two protons and two neutrons)
   c) atomic number = 5 (from five protons)
      mass number = 9 (from five protons and four neutrons)
   e) atomic number = 50 (from fifty protons)
      mass number = 119 (from fifty protons and sixty-nine neutrons)

2.13 In each case, the number of negatively-charged electrons must equal the number of positively charged protons.
   a) 2 electrons (to balance the charge of two protons)
   c) 5 electrons (to balance the charge of five protons)
   e) 50 electrons (to balance the charge of fifty protons)

2.14 In each case, the number of electrons must be equal to the number of protons in the nucleus, which is equal to the atomic number of the element.
   a) sulfur, S, atomic number = 16, so 16 electrons are present
   c) chromium, Cr, atomic number = 24, so 24 electrons are present

e) xenon, Xe, atomic number = 54, so 54 electrons are present

2.15 In each case, the superscript (upper number) is the mass number, which is equal to the sum of the number of protons plus the number of neutrons. The subscript (lower number) is equal to the atomic number, which is equal to both the number of electrons and the number of protons in the atoms. Thus, the superscript minus the subscript will equal the number of neutrons.
b) The atomic number = 10, so the atoms contain 10 protons and 10 electrons. The mass number = 22, so the atoms contain 22 - 10 or 12 neutrons.
d) The atomic number = 26, so the atoms contain 26 protons and 26 electrons. The mass number = 56, so the atoms contain 56 - 26 or 30 neutrons.
f) The atomic number = 47, so the atoms contain 47 protons and 47 electrons. The mass number = 109, so the atoms contain 109 - 47 or 62 neutrons.

2.16 In each case, the mass number or the superscript in the symbol is given as part of the name. The atomic number or subscript is obtained by matching the element name to its symbol, and locating the symbol with its atomic number in the periodic table located inside the front cover of the text.
a) The mass number is 110, and the atomic number of Cd from the periodic table is 48. $^{110}_{48}Cd$
c) The mass number is 235, and the atomic number of U from the periodic table is 92. $^{235}_{92}U$
e) The mass number is 14, and the atomic number of C from the periodic table is 6. $^{14}_{6}C$

2.17 a) The mass number follows the element name, so beryllium-10 and boron-10 have the same mass number.
c) The number of protons is the same as the atomic number that is obtained from the periodic table. The mass number is the total of protons + neutrons, so we are looking for elements with an atomic number that is one-half the given mass number.
Helium-4 (atomic number = 2), nitrogen-14 (atomic number = 7) and boron-10 (atomic number = 5) all meet this condition.

2.18 b) Because the elements contain only one isotope, the atomic weight in the periodic table will be the mass in amu of the atom's nucleus. The mass contributed by the electrons is negligible. When rounded to three significant figures, the masses are: Be, 9.01 amu; Na, 23.0 amu; Al, 27.0 amu.

2.19 In each case, the atomic weight will be calculated as

$$\text{atomic weight} = \frac{(\%)(mass) + (\%)(mass) + etc.}{100}$$

a) atomic weight

$$= \frac{(7.42\%)(6.0151\ amu) + (92.58\%)(7.0160\ amu)}{100}$$

$$= \frac{44.632\ amu + 649.541\ amu}{100} = \frac{694.173}{100}$$

$= 6.94$ amu. This agrees well with the periodic table value of 6.941 amu.

c) Atomic weight

$$= \frac{(92.21\%)(27.9769\ amu) + (4.70\%)(28.9765) + (3.09\%)(29.9738\ amu)}{100}$$

$$= \frac{2579.750\ amu + 136.190\ amu + 92.619\ amu}{100}$$

$$= \frac{2808.559\ amu}{100} = 28.1\ amu$$

This agrees well with the periodic table value of 28.09 amu.

e) Atomic weight =

$$\frac{(1.48\%)(203.973\ amu) + (23.6\%)(205.975\ amu)}{100}$$

$$\frac{+ (22.6\%)(206.976\ amu) + 52.3\% (207.977\ amu)}{100}$$

$$= \frac{301.880 \; amu + 4861.010 \; amu + 4677.658 \; amu + 10877.197 \; amu}{100}$$

$$= \frac{20717.745}{100} \; amu = 207 \; amu$$

This agrees well with the periodic table value of 207.2 amu.

2.20 According to their atomic weights, 12.0 g of carbon and 31.0 g of phosphorus would contain the same number of atoms. Thus, 1/2 as much or 6.0 g of carbon and 15.5 g of phosphorus would each contain the same number of atoms. This leads to the conclusion that 1/10 of 6.0 g of carbon or 0.60 g, and 1/10 of 15.5 g of phosphorus or 1.6 g would contain the same number of atoms.

2.22 A 100.0 g sample would contain 45.2 g of Al and 54.8 g of Zn. The ratio of the atomic weight of Zn to that of Al is 65.4/27.0 or 2.42/1 or 2.42:1. The ratio of the mass of Zn to Al in the mixture is 54.8/45.2 or 1.21/1 or 1.2:1. If the atoms were present in equal numbers, the ratio would be the same as that of the atomic weight, 2.42:1. Since the ratio is only half that, we can conclude that only half as many Zn atoms are present as Al atoms. Or, there are twice as many Al atoms as there are Zn atoms in the mixture.

2.23 In each case, the answer is based on the idea that 1 mole of an element is defined as the amount that would have a mass in grams equal to the atomic weight of the element. This amount contains 6.02×10^{23} atoms for any element.

2.24 a) $50.0 \; g \; K \times \dfrac{1 \; mol \; K}{39.1 \; g \; K} = 1.28 \; mol \; K$

c) $200 \; g \; Sn \times \dfrac{6.02 \times 10^{23} \; Sn \; atoms}{118.7 \; g \; Sn} = 1.01 \times 10^{24} \; Sn \; atoms$

e) $2.75 \; mol \; C \times \dfrac{12.0 \; g \; C}{1 \; mol \; C} = 33.0 \; g \; C$

2.25 a) $1.39 \text{ g Li} \times \dfrac{1 \text{ mol Li}}{6.94 \text{ g Li}} = 0.200 \text{ mol Li}$

c) The atoms react in a 1:1 ratio, so they would react in a 1:1 mole ratio. Thus, 0.200 mol of I would react.

$0.200 \text{ mol I} \times \dfrac{6.02 \times 10^{23} \text{ I atoms}}{1 \text{ mol I}} = 1.20 \times 10^{23} \text{ I atoms}$

e) $0.200 \text{ mol I} \times \dfrac{126.9 \text{ g I}}{1 \text{ mol I}} = 2.54 \times 10^{1} = 25.4 \text{ g I}$

2.26 The text answer is the explanation

2.27 b) $1.00 \text{ mol NH}_3 \times \dfrac{14.0 \text{ g N}}{1 \text{ mol NH}_3} = 14.0 \text{ g N}$

d) $0.50 \text{ mol } C_4H_{10}O \times \dfrac{10 \text{ mol H atoms}}{1 \text{ mol } C_4H_{10}O} = 5.0 \text{ mol H atoms}$

f) $2.00 \text{ mol } C_6H_7N \times \dfrac{7.0 \text{ g H}}{1 \text{ mol } C_6H_7N} = 14.0 \text{ g H}$

h) $1 \text{ mol } C_2H_3O_2F \times \dfrac{12.04 \times 10^{23} \text{ C atoms}}{1 \text{ mol } C_2H_3O_2} = 12.04 \times 10^{23} \text{ C atoms}$

$12.04 \times 10^{23} \text{ C atoms} \times \dfrac{1 \text{ mol } C_4H_{10}O}{4 \times 6.02 \times 10^{23} \text{ C atoms}} = 0.50 \text{ mol } C_4H_{10}O$

2.28 Consider one mole of gas in each case. Calculate the molecular weight of each gas, and the mass of oxygen contained in one mole. Then use the formula

$\% \text{ O} = \dfrac{\text{mass O in 1 mole}}{\text{mass of 1 mole}} \times 100.$

26 CHAPTER 2

N_2O: mass of N in one mole = 2 × 14.0 g = 28.0 g
mass of O in one mole = 1 × 16.0 g = 16.0 g
Total mass of one mole = 44.0 g

$$\% \; O = \frac{16.0 \text{ g}}{44.0 \text{ g}} \times 100 = 36.4\%$$

NO: mass of N in one mole = 1 × 14.0 g = 14.0 g
mass of O in one mole = 1 × 16.0 g = 16.0 g
Total mass of one mole = 30.0 g

$$\% \; O = \frac{16.0 \text{ g}}{30.0 \text{ g}} \times 100 = 53.3\%$$

NO_2: mass of N in one mole = 1 × 14.0 g = 14.0 g
mass of O in one mole = 2 × 16.0 g = 32.0 g
Total mass of one mole = 46.0 g

$$\% \; O = \frac{32.0 \text{ g}}{46.0 \text{ g}} \times 100 = 69.6\%$$

N_2O_4: mass of N in one mole = 2 × 14.0 g = 28.0 g
mass of O in one mole = 4 × 16.0 g = 32.0 g
Total mass of one mole = 92.0 g

$$\% \; O = \frac{64.0 \text{ g}}{92.0 \text{ g}} \times 100 = 69.6\%$$

2.29 a) $70.0 \text{ g } C_6H_5NO_3 \times \dfrac{14.0 \text{ g } N}{139.0 \text{ g } C_6H_5NO_3} = 7.05 \text{ g } N$

c) $9.00 \times 10^{22} \; C_6H_5NO_3 \text{ molecules} \times \dfrac{6 \times 6.02 \times 10^{23} \text{ C atoms}}{6.02 \times 10^{23} \; C_6H_5NO_3 \text{ molecules}}$

$$= 5.40 \times 10^{23} \text{ C atoms}$$

e) $18.0 \cancel{g\ N} \times \dfrac{6.02 \times 10^{23}\ N\ atoms}{72.0\ \cancel{g\ N}} = 1.51 \times 10^{23}\ N\ atoms$

2.31 Fe_3O_4: mass of Fe in one mole = 3 x 55.85 g = 167.6 g
    mass of O in one mole = 4 x 16.00 g = 64.0 g
    Total mass of one mole   = 231.6 g

$\%\ Fe = \dfrac{167.6\ \cancel{g}}{231.6\ \cancel{g}} \times 100 = 72.4\%$

Fe_2O_3: mass of Fe in one mole = 2 x 55.85 g = 111.7
    mass of O in one mole = 3 x 16.00 g = 48.00 g
    Total mass of one mole   = 159.7 g

$\%\ Fe = \dfrac{111.7\ \cancel{g}}{159.7\ \cancel{g}} \times 100 = 69.9\%$

SELF-TEST QUESTIONS

Multiple Choice

1. Which of the following is an incorrect symbol for an element?
   a) Ce
   b) Au
   c) K
   d) CR

2. Which of the following is an incorrect formula for a compound?
   a) CO_2
   b) CO_1
   c) N_2O
   d) NO_2

3. Three objects have masses of 3.2 g, 1.6 g, and 0.80 g. What is the relative mass of the 3.2 g object compared to the others?
   a) 4.0
   b) 2.0
   c) 0.50
   d) 0.25

28 CHAPTER 2

4. Suppose the atomic weights of the elements were assigned in such a way that the atomic weight of helium, He, was 1.00 amu. What would be the atomic weight of oxygen, O, in amu on this scale?
   a) 16.0
   b) 8.00
   c) 4.00
   d) 0.250

5. What is the molecular weight of phosphoric acid, H_3PO_4, in amu?
   a) 48.0
   b) 50.0
   c) 96.0
   d) 98.0

6. How many neutrons are there in the nucleus of a potassium-39 atom?
   a) 1
   b) 19
   c) 20
   d) 39

7. What is the mass in grams of 1.00 mole of chlorine molecules, Cl_2?
   a) 6.02×10^{23}
   b) 70.9
   c) 35.5
   d) 1.18×10^{-22}

8. Calculate the weight percent of sulfur, S, in SO_2
   a) 50.1
   b) 33.3
   c) 66.7
   d) 25.0

Matching

Match the numbers given as responses to the following.

9. The number of moles of oxygen atoms in 2 moles of NO_2

10. The number of moles of NH_3 that contains 3 moles of nitrogen atoms

11. The number of moles of nitrogen atoms in one-half mole of N_2O_5

12. The number of moles of electrons in one mole of helium atoms

13. The number of moles of neutrons in one mole of ^3_1H

a) 1
b) 2
c) 3
d) 4

True-False

14. In some instances, two different elements are represented by the same symbol.

15. The mass of a single atom of silicon, Si, is 28.1 grams.

16. All isotopes of a specific element have the same atomic number.

17. One mole of water molecules, H$_2$O, contains two moles of hydrogen atoms, H.

18. 1.00 mol of sulfur, S, contains the same number of atoms as 14.0 grams of nitrogen, N.

ANSWERS TO PROGRAMMED REVIEW

2.1   a) name   b) symbol   c) symbols   d) capital letter
      e) capital letter   f) small letter   g) formula   h) symbol

2.2   a) relative masses   b) atomic weights   c) molecular weights
      d) atomic mass units

2.3   a) nucleus   b) protons   c) neutrons   d) neutrons
      e) one   f) protons   g) one   h) electrons   i) 1/1836

2.4   a) protons   b) mass number   c) isotopes

2.5   a) mass number   b) average

2.6   a) 79.90

2.7   a) mole   b) 6.02 x 10^{23} c) 83.80

2.8   a) three   b) 44.02   c) 28.02   d) 16.00

ANSWERS TO SELF-TEST QUESTIONS

1.	d	7.	b	13.	b
2.	b	8.	a	14.	F
3.	a	9.	d	15.	F
4.	c	10.	c	16.	T
5.	d	11.	a	17.	T
6.	c	12.	b	18.	T

CHAPTER 3

Electronic and Nuclear Characteristics

PROGRAMMED REVIEW

Section 3.1 The Periodic Law and Table

According to the (a) _____ _____, the properties of elements arranged by increasing atomic numbers repeat at regular intervals. In a modern (b) _____ _____, elements with similar properties are found in vertical columns called (c) _____ or (d) _____. The (e) _____ rows in the periodic table are called periods.

Section 3.2 Electronic Arrangements in Atoms

Neils Bohr improved our understanding of atomic structure by suggesting a modification to the (a) _____ _____ model proposed by Ernest Rutherford. Bohr suggested that (b) _____ could occupy only (c) _____ located (d) _____ _____ from the nucleus. He also theorized that electrons changed (e) _____ only by absorbing or releasing energy. However, additional research suggested that electrons did not follow specific paths around the nucleus, but instead moved in specific (f) _____ of space called (g) _____ _____. Atomic orbitals occur in groups called (h) _____ which are designated by the same number and letter used to designate the (i) _____ within the subshell. Subshells occur in groups called (j) _____.

Section 3.3 The Shell Model and Elemental Properties

Similar elemental properties result when elements have identical numbers of electrons in the (a) _____ _____ of their atoms. Elements in groups IIA(2), VA(15) and VIIA(17) have respectively the following numbers of electrons in their valence shells: (b) _____, _____ and _____. With the exception of helium, all noble gases have (c) _____ electrons in their valence shells.

Section 3.4 Electronic Configurations

The detailed arrangements of electrons in atoms are called (a) _____ _____. As electrons are added to atoms, they will occupy the subshell of (b) _____ _____ that is available. According to (c) _____ _____, electrons will not pair up in orbitals as long as empty orbitals of the same energy are available. Two electrons that occupy the same orbital must be spinning in opposite (d) _____, in compliance with the (e) _____ _____ _____.

Section 3.5 Other Electronic Configuration Representations

An electronic configuration that ends with a completely filled p subshell is called a (a) _____ _____ _____. Noble gas configurations can be used to write shortened electronic configurations by letting the noble gas symbol enclosed in brackets represent the (b) _____ found in the noble gas configuration. In an (c) _____ _____ _____ or (d) _____ _____ for an atom or ion, the valence electrons are represented by dots arranged around the elemental symbol.

Section 3.6 Another Look at the Periodic Table

The last or highest energy electron in an atom is called the (a) _____ _____. Elements found in the s and p areas of the periodic table are called (b) _____ _____. The (c) _____ area of the periodic table contains (d) _____ _____. Elements classified as (e) _____ are found in the left two-thirds of the periodic table and in the f area. Elements classified as (f) _____ form a diagonal separation zone between metals and nonmetals.

Section 3.7 Radioactive Nuclei

(a) _____ nuclei undergo spontaneous changes and emit energy. The emission of radiation by unstable nuclei is called (b) _____ _____. The common types of emitted radiation are (c) _____ particles, (d) _____ particles and (e) _____ rays.

Section 3.8 Equations for Nuclear Reactions

In equations used to represent nuclear reactions, all particles are designated by a (a) _____, a (b) _____ number and an (c) _____ number. A nuclear equation is balanced when the sums of the (d) _____ numbers and (e) _____ numbers are the same on both sides of the equation. A nucleus produced by radioactive decay is called a (f) _____. When a nucleus emits a (g) _____, a nuclear proton is changed to a neutron.

Section 3.9 Isotope Half-life

A (a) _____ life is the time required for (b) _____ the atoms in a sample to undergo radioactive decay. Radioisotopes found in nature have (c) _____ half-lives, are (d) _____ of the decay of long-lived isotopes, or are produced by natural processes such as cosmic ray bombardment. After two half-lives have passed, the fraction of original atoms remaining is (e) _____.

Section 3.10 Health Effects of Radiation

The greatest danger of radiation to living organisms results from the ability of radiation to generate (a) _____ or (b) _____ _____. The condition associated with short-term exposure to intense radiation is called (c) _____ _____. Two protections against radiation are (d) _____ and (e) _____.

Section 3.11 Measurement Units for Radiation

Two types of units used to describe quantities of radiation are (a) _____ units that indicate the activity of the source, and (b) _____ units that are related to the tissue damage caused by the radiation. Two physical units are the (c) _____ and the (d) _____. A biological unit used with x-rays and gamma rays is the (e) _____. Both the (f) _____ and (g) _____ describe the effects of radiation in terms of the energy transferred to tissue. A (h) _____ is a biological unit that accounts for health differences in various types of radiation. Three devices that are used to detect radiation are (i) _____ badges, (j) _____ counters and (k) _____-_____ tubes.

Section 3.12 Medical Uses of Radioisotopes

In diagnostic applications, radioisotopes are used as (a) _____. Radioisotopes used for this purpose ideally should have short (b) _____-_____, but not too short since they must be prepared and administered conveniently. These isotopes also should produce (c) _____ that are nontoxic and, ideally, (d) _____ toward further radioactive decay. The radioisotopes used diagnostically should preferably give off (e) _____ radiation, and they should be absorbed by tissue to form (f) _____ _____ or rejected by tissue to form (g) _____ _____. Radioisotopes used therapeutically should ideally emit (h) _____ or (i) _____ radiation. Their (j) _____-_____ should be long enough to allow therapy to be accomplished. Their daughters should be non (k) _____ and give off little or no (l) _____. Therapeutic radioisotopes should be (m) _____ by the body in the target tissue.

Section 3.13 Nonmedical Uses of Radioisotopes

Radioisotopes are used as (a) _____ in some nonmedical applications such as following the path of a compound in a process, indicating the boundary between products in a (b) _____, and measuring the effectiveness of (c) _____. Some radioisotopes are used to determine the ages of artifacts or minerals in a process called (d) _____ _____.

34 CHAPTER 3

SOLUTIONS TO EXERCISES ANSWERED IN THE TEXT

3.1 In each case, the group is the vertical column in which the element is found, and the period is the horizontal row in which it is found. The element in question is located in the square where the two cross in the periodic table.
b) Element number 22 is titanium, Ti. It belongs to group IVB (4) and period 4.
d) Tin, Sn, is element number 50. It belongs to group IVA (14) and period 5.
f) Argon, Ar, is element number 18. It belongs to the group of noble gases (18) and period 3.

3.2 The element in question will be located in the square where the given group and period cross.
a) sulfur, S   c) silicon, Si   e) radon, Rn

3.3 a) Group VIII (8, 9, 10) is three groups in the new system for designating groups. It contains eleven elements.
c) Group IIA (2) contains six elements while group VIA (16) contains five.

3.4 b) According to the Bohr theory, energy must be put in to move an electron to an orbit farther from the nucleus. Thus, an electron in the orbit located farther from the nucleus would have the higher energy.

3.7 a) The answer is 2, because all orbitals regardless of their designation can contain a maximum of two electrons.
c) The answer is 2 because all orbitals regardless of their designation can contain a maximum of two electrons.
e) The answer is 2 because the first shell ($n = 1$) contains only one orbital, and each orbital can contain a maximum of two electrons.

3.8 The third shell ($n = 3$) contains three subshells; a $3s$, a $3p$ and a $3d$. An s subshell contains only a single orbital. p subshells contain 3 orbitals, and d subshells contain 5 orbitals. Thus, the shell contains $1 + 3 + 5 = 9$ orbitals.

3.10 The fourth shell ($n = 4$) contains an s subshell ($4s$), a p subshell ($4p$), a d subshell ($4d$) and an f subshell ($4f$). The subshells respectively contain 1, 3, 5 and 7 orbitals. Each orbital can contain a maximum of two electrons, so the subshells can contain respectively 2, 6, 10 and 14 electrons. The maximum total number of electrons in the shell is thus $2 + 6 + 10 + 14 = 32$.

3.11 b) Pb is in group IVA (14) and thus contains 4 valence-shell electrons.
d) Bismuth, Bi, is in group VA (15) and thus contains 5 valence-shell electrons.
f) Element number 49 is indium, In. It belongs to group IIIA (13) and so contains 3 valence-shell electrons.

Electronic and Nuclear Characteristics 35

3.12 Cesium, Cs, is a period 6 element that is also in group IA (1), the same group as sodium. It has one valence-shell electron just as sodium has, and so would have properties similar to sodium.

3.14 The number of unpaired electrons is easily determined by putting the valence-shell electrons into their orbitals following Hund's rule.
   b) K, number 19, is in group IA (1). It has one valence electron among its total 19 electrons, $1s^2 2s^2 2p^6 3s^2 3p^6 4s^1$. The valence-shell electron is the only electron in the 4s orbital and so is unpaired.
   d) Selenium, Se, is element number 34. It belongs to group VIA and so has 6 valence-shell electrons among its total 34 electrons, $1s^2 2s^2 2p^6 3s^2 3p^6 4s^2 3d^{10} 4p^4$. The valence-shell electrons are the $4s^2$ and $4p^4$ electrons. The $4s^2$ orbital contains its maximum of 2 electrons which are paired. The four 4p electrons will occupy the three 4p orbitals as follows: ↓↑ ↓ ↓ . Thus, two electrons are unpaired.
   f) Element 36, Kr, is a noble gas (18) element. Since the noble gases have completely filled subshells, all orbitals contain 2 electrons, and so all electrons are paired. No unpaired electrons are present. $1s^2 2s^2 2p^6 3s^2 3p^6 4s^2 3d^{10} 4p^6$
   h) Sb, element number 51 belongs to group VA (15) and so has five valence-shell electrons among its 51 total electrons: $1s^2 2s^2 2p^6 3s^2 3p^6 4s^2 3d^{10} 4p^6 5s^2 4d^{10} 5p^3$. The valence-shell electrons are the $5s^2$ and $5p^3$. The $5s^2$ subshell is filled, so all electrons are paired. The $5p^3$ electrons occupy the 5p orbitals as follows: ↓ ↓ ↓ . Thus, three electrons are unpaired.

3.15 a) Mg: $1s^2 2s^2 2p^6 3s^2$: The total number of s electrons is obtained by counting. It is 6.
   c) Al: $1s^2 2s^2 2p^6 3s^2 3p^1$: The filled subshells are 1s, 2s, 2p, and 3s. Remember, s subshells contain one orbital and a maximum of 2 electrons, while p subshells contain three orbitals and a maximum of 6 electrons.
   e) S: $1s^2 2s^2 2p^6 3s^2 3p^4$: The 2p subshell is filled, so all those p electrons are paired. The four 3p electrons fill the subshell as follows: ↓↑ ↓ ↓ . Thus, two of the 3p electrons are unpaired. Their number designation is 3.

3.16 b) The element that satisfies the criterion must belong to period 3 because of the number designation, and group IA(1) because only group IA(1) elements have single and thus unpaired s electrons. The element is sodium, Na.
   d) The element that satisfies the criterion must belong to period 5 because the number designation of d electrons is one less than the period number. The element must be the third element to receive d electrons in that period because it contains only three d electrons. The element is niobium, Nb.
   f) An element that satisfies the criterion must belong to period 5 because of the number designation. It must contain only one or seven p electrons in order to have a single p electron unpaired: either ↓ __ __ or ↓↑ ↓↑ ↓ . The elements that satisfy the criterion are indium, In, and iodine, I.

h) An element that satisfies the criterion must belong to period 6 because of the number designation. It must contain only three p electrons because a p subshell can contain a maximum of six electrons, so 3 electrons will half-fill the subshell. The element is bismuth, Bi.

j) The element that satisfies the criterion must be an inner-transition element. It will contain 7 $4f$ electrons, since an f subshell can contain a maximum of 14 electrons. The element is gadolinium, Gd.

3.17 The noble-gas symbol used in each case will be that of the noble gas that precedes the period in which the element occurs.
 a) Li: [He] $2s^1$
 c) This element will contain 15 electrons and will be element number 15 or phosphorus, P: [Ne] $3s^2 3p^3$
 e) Element 53 is iodine, I: [Kr] $5s^2 4d^{10} 5p^5$

3.19 All elements that follow Kr and precede Xe would have Kr as a part of their electronic configuration written using noble gas symbols. This includes the 17 elements numbered from 37 through 53.

3.20 In each case the valence-shell electrons will be represented by a dot around the symbol for the element. The number of valence-shell electrons is the same as the group number to which the element belongs.
 a) Lithium is in group IA(1) and contains 1 valence-shell electron: Li·
 c) Potassium is in group IA(1) and contains 1 valence-shell electron: K·
 e) Carbon is in group IVA(14) and contains 4 valence-shell electrons:

$$\cdot \overset{\cdot}{\underset{\cdot}{C}} \cdot \quad \text{or} \quad : \overset{\cdot}{C} \cdot \quad \text{or} \quad : C :$$

3.21 b) A group IVA(14) element will contain 4 valence-shell electrons:

$$\cdot \overset{\cdot}{\underset{\cdot}{E}} \cdot \quad \text{or} \quad : \overset{\cdot}{E} \cdot \quad \text{or} \quad : E :$$

3.22 In each case, the distinguishing electron is the last electron in the written electronic configuration.
 a) Element number 5 has 5 electrons: $1s^2 2s^2 2p^1$. Thus, the distinguishing electron is a p electron.
 c) Ge is element number 32: [Ar] $4s^2 3d^{10} 4p^2$. Thus, the distinguishing electron is a p electron.

e) Zirconium, Zr, is element number 40: [Kr] $5s^2 4d^2$. Thus, the distinguishing electron is a *d* electron.

3.23 In each case the element's number is matched to the numbers given in Figure 3.11.
a) Representative element
c) Representative element
e) Transition element

3.24 In each case, determine the element's number, then compare the number to the numbers and categories given in Figure 3.11.
b) Cesium, Cs, is element number 55: Representative element
d) Samarium, Sm, is element number 62: Inner-transition element
f) Curium, Cm, is element number 96: Inner-transition element
h) Helium, He, is element number 2: Noble gas element

3.25 In each case, determine the element's number, then compare the number to the numbers and categories given in Figure 3.13.
b) Platinum, Pt, is element number 78: Metal
d) Element number 67 is holmium, Ho: Metal
f) Arsenic, As, is element number 33: Metalloid
h) S is sulfur, element number 16: Nonmetal

3.27 The answers are obtained by comparison to the information given in Table 3.3.

3.28 The composition of the particles is obtained from information given in Table 3.3.
b) An alpha particle is composed of 2 protons and 2 neutrons.

3.30 Explanations are given in the answers of appendix B of the text. Remember, the mass number is the sum of the number of protons plus the number of neutrons in a nucleus.

3.31 In each case, the A represents the mass number, or the sum of the number of protons plus neutrons in the nucleus. The Z represents the atomic number or number of protons in the nucleus.
b) Chromium is element number 24, and so contains 24 protons in the nucleus. The number of neutrons is given as 26, so Z = 24, and A = 24 + 26 = 50.
d) The element that belongs to period 5 and group IB(11) is silver, Ag. It is element number 47 and so has 47 protons in the nucleus. The mass number is given as 109. Thus, Z = 47, and A = 109.
f) The one proton means Z = 1, and the one proton plus two neutrons means A = 3.

38   CHAPTER 3

3.32   In each case, the sum of the mass numbers, A, and the atomic numbers Z (or charge) must be equal on each side of the equation.
b) The required particle must have A = 206 and Z = 81. This corresponds to thallium-206 with the symbol given as the answer.
d) The required particle must have A = 44 and Z = 21. This corresponds to scandium-44 with the symbol given as the answer.
f) The required particle must have A = 0 and Z = 1. This corresponds to positron with the symbol given as the answer.
h) The required particle must have A = 69 and Z = 31. This corresponds to gallium-69 with the symbol given as the answer.

3.33   In each case the sum of the mass numbers, A, and the atomic numbers, Z (or charge) must be equal on each side of the balanced equation. The symbol for the missing product is obtained by determining the atomic number of the product and comparing it to the symbols in the periodic table.
b) The sum of the A values on the left (the iron-55 plus the captured electron) is 55. The sum of the Z values on the left (the iron-55 plus the captured electron) is 25. Thus, the product will have A = 55 and Z = 25. This corresponds to manganese-55 with the symbol given as the answer.
d) The A value on the left is 190, and the Z value is 78. With an alpha particle (A = 4 and Z = 2) as one of the products, the other product must have A = 186 and Z = 76. This corresponds to osmium-186 with the symbol given as the answer.
f) The A value on the left is 67, and the Z value is 31. The daughter is zinc-67 with an A value of 67 and a Z value of 30. Thus, the A value of the missing particle is 0, and the Z value is 1. This corresponds to a positron with the symbol given as the answer.
h) The A value on the left is 154, and the Z value is 66, The daughter is gadolinium-150 with an A value of 150 and a Z value of 64. Thus, the A value of the missing particle is 4, and the Z value is 2. This corresponds to an alpha particle with the symbol given as the answer.

3.35   The half-life for any material is the time it takes for a sample to be reduced to one-half the value it had when the timing was begun. Thus, in each case the time it takes to reduce the quantity to one-half it's original value is the half-life.

3.37   A remaining fraction of 1/128 corresponds to the passage of 7 half-lives (½ x ½ x ½ x ½ x ½ x ½ x ½ = 1/128). Seven half-lives of 92 years = 7 x 92 years = 644 years. Thus, the archaeologist made the discovery 644 years after the building was constructed, or in 1980 + 644 or 2624.

3.39   Ten hours is equal to 4 half-lives. Thus, the fraction of the original amount that would remain would be ½ x ½ x ½ x ½ = 1/16. The amount remaining would then be 50.0 mg/16 = 3.13 mg.

3.41 Equation 3.2 is used: $I_x/I_y = d_y^2/d_x^2$. In the problem given, let $I_x = 90$ units when $d_x = 10$ feet, and $I_y = 5$ units when $d_y = $? Then, $d_y^2 = I_x/I_y \times d_x^2 = 90/5 \times (10)^2 = 90/5 \times 100 = 1800$ ft^2

$$d_y = \sqrt{1800 \text{ ft}^2} = 42.4 \text{ ft.}$$

3.43 Explanation is given as the answer in Appendix B.

3.45 Explanation is given as the answer in Appendix B.

3.47 Explanation is given as the answer in Appendix B.

3.49 Explanation is given as the answer in Appendix B.

3.51 Explanation is given as the answer in Appendix B.

SELF-TEST QUESTIONS

Multiple Choice

1. Elements in the same group
   a) have similar chemical properties
   b) are called isotopes
   c) have consecutive atomic numbers
   d) constitute a period of elements

2. Which element with the following atomic numbers should have properties similar to those of oxygen (element number 8)?
   a) 15   b) 4   c) 2   d) 34

3. Which of the following elements is found in period 3 of the periodic table?
   a) Al
   b) B
   c) Ga
   d) more than one response is correct

4. The electronic configuration for an element containing 15 protons would be
   a) $1s^2 2s^2 2p^6 3s^2 3p^6$
   b) $1s^2 2s^2 2p^6 3s^2 4p^3$
   c) $1s^2 2s^2 2p^6 3s^2 3p^6 4s^2$
   d) $1s^2 2s^2 2p^6 3s^2 3p^3$

5. Which of the following is a true statement for an electronic configuration of $1s^2 2s^2 2p^6 3s^2 3p^6$?
   a) there are 6 electrons in the $3p$ orbital
   b) there are 6 electrons in the $3p$ subshell
   c) there are 6 electrons in the $3p$ shell
   d) more than one response is correct

6. The maximum number of electrons which may occupy a $4d$ orbital is
   a) 10   b) 4   c) 2   d) 8

7. How many unpaired electrons are found in titanium (element number 22)?
   a) 1   b) 2   c) 3   d) 4

8. Which element has the electronic configuration $1s^2 2s^2 2p^6 3s^2 3p^6$?
   a) Ne   b) Ar   c) K   d) Kr

9. The element with unpaired electrons in a d subshell is element number
   a) 33   b) 27   c) 63   d) 53

10. Which of the following elements has an electronic configuration ending in $4d^7$?
    a) Co   b) Rh   c) Ir   d) more than one is correct

11. How many electrons are found in the valence shell of oxygen (element number 8)?
    a) 4   b) 3   c) 5   d) 6

12. The distinguishing electronic configuration of np^4 is characteristic of which group in the periodic table?
    a) IIA (2)
    b) IVA (14)
    c) VIA (16)
    d) noble gases

13. The element with the electronic configuration $1s^2 2s^2 2p^3$ will be found in
    a) period 1, group VIA (16)
    b) period 2, group VA (15)
    c) period VA (15), group 2
    d) period IIIA (13), group VA (15)

14. How many electrons would be contained in the valence shell of S in group VIA (16)?
    a) 6   b) 2   c) 4   d) 16

15. The three common types of radiation emitted by naturally radioactive elements are
    a) electrons, protons and neutrons
    b) x-rays, gamma rays and protons
    c) alpha rays, beta rays and neutrons
    d) alpha rays, beta rays and gamma rays

16. Which of the following types of radiation is composed of particles which carry a +2 charge?
    a) alpha b) beta c) gamma d) neutrons

17. Which of the following types of radiation is not composed of particles?
    a) alpha b) beta c) gamma d) neutrons

18. After four half-lives have elapsed, the amount of a radioactive sample which has not decayed is
    a) 40% of the original amount
    b) ¼ of the original amount
    c) ⅛ of the original amount
    d) 1/16 of the original amount

19. If ⅛ of an isotope sample is present after 22 days, what is the half-life of the isotope?
    a) 10 days
    b) 5 days
    c) 2¾ days
    d) 7⅓ days

20. By doubling the distance between yourself and a source of radiation how is the intensity of the radiation getting to you changed?
    a) it is ½ as great
    b) it is ⅓ as great
    c) it is ¼ as great
    d) it is ⅛ as great

21. Which of the following would be the most convenient unit to use when determining the total dose of radiation received by an individual who was exposed to several different types of radiation?
    a) Roentgen b) rad c) gray d) rem

True-False

22. Elements 20 and 21 are in the same period of the periodic table.

23. Hund's rule states that electrons within a subshell remain unpaired if possible.

24. A 2p and a 3p subshell contain the same number of orbitals.

25. The maximum number of electrons an orbital may contain does not vary with the type of orbital.

26. The distinguishing electron in Br is found in a *p* orbital.

27. Radioactive tracers are useful in both medical and nonmedical applications.

28. Radioactive isotopes are not taken into the body during medical uses.

29. A Curie is a physical measurement of the quantity of radiation.

30. The rem is a biological radiation measurement unit.

Matching

From the list on the right, choose a response that is consistent with the description on the left as far as electronic configurations are concerned. You may use a response more than once.

31. a noble gas

32. Mg

33. a transition element

34. an element just completing the filling of the shell where n = 2

a) $1s^2 2s^2 2p^6 3s^2$
b) $1s^2 2s^2 2p^6$
c) $1s^2 2s^2 2p^6 3s^2 3p^6 4s^2 3d^3$
d) $1s^2 2s^2 2p^4$

Match each of the categories on the right with an element on the left.

35. neon (Ne)

36. phosphorus (P)

37. calcium (Ca)

38. element number 47

39. element number 82

40. top element of group VA (15)

a) a representative metal
b) a noble gas nonmetal
c) a nonmetal but not a noble gas
d) a transition metal

For each of the nuclear equations on the left choose the correct identity of X from the choices on the right.

41. $^{13}_{7}N \rightarrow \,^{13}_{6}C + X$

42. $^{27}_{13}Al + \,^{2}_{1}H \rightarrow \,^{25}_{12}Mg + X$

43. $^{9}_{4}Be + \,^{4}_{2}He \rightarrow \,^{12}_{6}C + X$

a) beta, $^{0}_{-1}\beta$
b) neutron, $^{1}_{0}n$
c) positron, $^{0}_{1}\beta$
d) alpha, $^{4}_{2}\alpha$

ANSWERS TO PROGRAMMED REVIEW

3.1 a) periodic law   b) periodic table   c) groups   d) families
     e) horizontal

3.2 a) solar system   b) electrons   c) orbits   d) specific distances
     e) orbits   f) volumes   g) atomic orbitals   h) subshells
     i) orbitals   j) shells

3.3 a) valence shells   b) two, five and seven   c) eight

3.4 a) electronic configurations   b) lowest energy   c) Hund's rule
     d) directions   e) Pauli exclusion principle

3.5 a) noble gas configuration   b) electrons   c) electron dot formula
     d) Lewis structure

3.6 a) distinguishing electron   b) representative elements   c) *d*
     d) transition elements   e) metals   f) metalloids

3.7 a) radioactive   b) radioactive decay   c) alpha   d) beta
     e) gamma

3.8 a) symbol   b) mass   c) atomic   d) mass   e) atomic
     f) daughter   g) positron

3.9 a) half   b) half   c) long   d) daughters   e) ¼

3.10 a) radicals   b) free radicals   c) radiation sickness
     d) shielding   e) distance

3.11 a) physical   b) biological   c) Curie   d) Becquerel
     e) Roentgen   f) rad   g) gray   h) rem   i) film
     j) scintillation   k) Geiger-Müller

3.12 a) tracers   b) half-lives   c) daughters   d) stable
     e) gamma   f) hot spots   g) cold spots   h) alpha
     i) beta   j) half-lives   k) toxic   l) radiation
     m) concentrated

3.13 a) tracers   b) pipeline   c) lubricants   d) radioactive dating

ANSWERS TO SELF-TEST QUESTIONS

1.	a	16.	a	30.	T
2.	d	17.	c	31.	b
3.	a	18.	d	32.	a
4.	d	19.	d	33.	c
5.	b	20.	c	34.	b
6.	c	21.	d	35.	b
7.	b	22.	T	36.	c
8.	b	23.	T	37.	a
9.	b	24.	T	38.	d
10.	b	25.	T	39.	a
11.	d	26.	T	40.	c
12.	c	27.	T	41.	c
13.	b	28.	F	42.	d
14.	a	29.	T	43.	b
15.	d				

CHAPTER 4

Electronic and Nuclear Characteristics

PROGRAMMED REVIEW

Section 4.1 Noble-gas Configurations

The electronic structure of noble gases represents a (a) _____ configuration. A noble gas configuration is characterized by (b) _____ electrons in the valence shell of helium and (c) _____ valence-shell electrons for other members of the group.

Section 4.2 Ionic Bonding

According to the (a) _____ _____, atoms tend to interact by achieving noble gas electronic configurations. Some atoms achieve noble gas configurations by transferring electrons and becoming charged atoms called (b) _____ _____. The attraction between oppositely charged ions is called an (c) _____ _____. During ionic bond formation, metals generally (d) _____ electrons and nonmetals generally (e) _____ them. Atoms and simple ions that have the same electronic confiugrations are said to be (f) _____ with each other.

Section 4.3 Ionic Compounds

Ionic compounds containing ions of only two elements are called a) _____ compounds. Formulas for ionic compounds do not represent the formulas of molecules, but only the simplest (b) _____ _____ of the ions in the compounds. The stable form of ionic compounds is a (c) _____ in which ions of opposite charge occupy (d) _____ sites in a crystal lattice.

Section 4.4 Covalent Bonding

Some elements combine by (a) _____ electrons rather than giving them up or accepting them. The net attractive force that results between atoms that share electrons is called a (b) _____ _____.

46 CHAPTER 4

Section 4.5 Molecular Shapes

The atoms of most molecules form distinct (a) _____-_____ shapes. The covalent bonds resulting from the sharing of two and three pair of electrons are referred to respectively as (b) _____ and _____ bonds. The shape of molecules can be predicted by applying the (c) _____ theory to the valence-shell electrons of the (d) _____ atoms of molecules.

Section 4.6 Covalent Molecules

Covalent bonds in which bonding electrons are shared equally are called (a) _____ covalent bonds. Differences in the (b) _____ of covalently bonded atoms cause bonding electrons to be shared unequally; such bonds are called (c) _____ covalent bonds. (d) _____ _____ result when polarized bonds create an unsymmetrical charge distribution in molecules.

Section 4.7 Polyatomic Ions

Covalently bonded groups of atoms that carry a net charge are called (a) _____ ions. With the exception of the ammonium ion, the common polyatomic ions are (b) _____ charged.

Section 4.8 Other Interparticle Forces

A solid in which lattice sites are occupied by atoms that are covalently bonded to each other is called a (a) _____ solid. The (b) _____ bond is the name given to the forces resulting from the attraction of positive (c) _____ to mobile electrons in a crystal lattice. The attractive forces between positive and negative ends of polar molecules are called (d) _____ forces. (e) _____ _____ results from the attractions of polar molecules in which hydrogen is covalently bonded to very electronegative elements. The weakest interparticle forces are called (f) _____ forces.

SOLUTIONS TO EXERCISES ANSWERED IN THE TEXT

4.1 The noble gas symbol used will be the one that precedes the period in which the element is found.
    b) Element 20, Ca, is in period 4 and is preceded by the noble gas Ar. [Ar] $4s^2$
    d) Phosphorus, element 15, is in period 3 and is preceded by the noble gas Ne. [Ne] $3s^23p^3$

Forces Between Particles 47

f) Element number 6, C, is in period 2 and is preceded by the noble gas He. [He] $2s^2 2p^2$

h) Cl, chlorine, is in period 3 and is preceded by the noble gas Ne. [Ne] $3s^2 3p^5$

4.2 The valence shell electrons will be shown as dots surrounding the symbol for the element.

b) Element 20, Ca, has 2 valence electrons: Ca·

d) Phosphorus, P, has 5 valence electrons: ·P·

f) Element 6, C, has 4 valence electrons: ·C·

h) Chlorine, Cl, has 7 valence electrons: :Cl·

4.3 Each element can achieve a noble gas configuration by either removing enough electrons to match the configuration of the noble gas that precedes it in the periodic table, or by adding enough electrons to match the configuration of the noble gas that follows it in the periodic table.
b) Ca: Remove 2 to match Ar, or add 16 to match Kr
d) P: Remove 5 to match Ne, or add 3 to match Ar
f) C: Remove 4 to match He, or add 4 to match Ne
h) Cl: Remove 7 to match Ne, or add 1 to match Ar

4.4 Each element will either lose or add electrons to achieve a noble gas configuration. Whichever process requires fewer electrons (the loss or the addition) is the one that will take place.
b) Al: Remove 3 to match Ne, or add 5 to match Ar. Three will be lost, and the Al atoms will become Al^{3+} ions. Al → Al^{3+} + 3e⁻
d) Element 34 is Se: Remove 16 to match Ar, or add 2 to match Kr. Two will be added, and the Se atoms will become Se^{2-} ions. Se + 2e⁻ → Se^{2-}
f) Oxygen is O: Remove 6 to match He, or add 2 to match Ne. Two will be added, and the O atoms will become O^{2-} ions. O + 2e⁻ → O^{2-}
h) I: Remove 17 to match Kr, or add 1 to match Xe. One will be added, and the I atoms will become I⁻ ions. I + e⁻ → I⁻

4.5 The word *isoelectronic* means that the electronic structure will be identical. All of the ions will have the electronic configuration of the noble gas that precedes them (positive ions) or follows them (negative ions) in the periodic table.

Electronic configuration of He, $1s^2$: Li⁺

48   CHAPTER 4

Electronic configuration of Ne, $1s^22s^22p^6$: F⁻, O²⁻, Na⁺, Ne
Electronic configuration of Ar, $1s^22s^22p^63s^23p^6$: P³⁻, Cl⁻
Electronic configuration of Kr, $1s^22s^22p^63s^23p^64s^23d^{10}4p^6$: Rb⁺, Kr, Se²⁻

4.6   b) Ba will lose 2 electrons to match Xe and become a Ba²⁺ ion. F will gain 1 electron to match Ne and will become an F⁻ ion. The Ba²⁺ and F⁻ ions will combine in a 1:2 ratio respectively in order to balance the charges of the ions in the final product.
d) Potassium, K, will lose 1 electron to match Ar and will become a K⁺ ion. Bromine, Br, will gain 1 electron to match Kr and will become a Br⁻ ion. The K⁺ and Br⁻ ions will combine in a 1:1 ratio respectively to balance the charges of the ions in the final product.
f) Element number 13, Al, will lose 3 electrons to match Ne and will become an Al³⁺ ion. Element number 35, Br, will gain 1 electron to match Kr and will become a Br⁻ ion. The Al³⁺ and Br⁻ ions will combine in a 1:3 ratio respectively to balance the charges of the ions in the final product.

4.7   Binary compounds are those that contain atoms of two different elements. Note that the formulas may contain more than two atoms as long as only two different elements are present.
b) OF₂ is binary even though three atoms are represented because only two elements (O and F) are present.
d) H₂S is binary even though three atoms are represented because only two elements (H and S) are present.
f) NaHCO₃ is not a binary compound because four elements are present(Na, H, C, and O).
h) Na₂O is binary even though three atoms are represented because only two elements (Na and O) are present.

4.8   In each case the metal, which appears first in the formula, is named first followed by the stem of the nonmetal with -ide added.
b) The metal is strontium and the nonmetal is chlorine. The name is strontium chloride.
d) The metal is lithium and the nonmetal is bromine. The name is lithium bromide.
f) The metal is sodium and the nonmetal is sulfur. The name is sodium sulfide.
h) The metal is cesium and the nonmetal is selenium. The name is cesium selenide.

4.9   Positive metal ions are simply named as the metal name followed by the word *ion*. Negative nonmetal ions are named by adding -ide to the stem of the name of the element.
a) Al³⁺ is a positive ion of the metal aluminum. The name is aluminum ion.
c) Mg²⁺ is a positive ion of the metal magnesium. The name is magnesium ion.
e) P³⁻ is a negative ion of the nonmetal phosphorus. The name is phosphide ion.

Forces Between Particles   49

g) I⁻ is a negative ion of the nonmetal iodine. The name is iodide ion.
i) N³⁻ is a negative ion of the nonmetal nitrogen. The name is nitride ion.

4.10   a) The chloride ion has a -1 charge, so the charge on the metal ion may be determined. In $CrCl_2$, the charge on the Cr must be +2 to match the -2 of the two Cl⁻ ions. Chromium (II) chloride. In $CrCl_3$, the charge on the Cr must be +3 to match the -3 charge of the three Cl⁻ ions. Chromium (III) chloride.
c) The oxygen ion has a -2 charge. In FeO the iron must have a +2 charge to match the -2 charge of the one O^{2-} ion. Iron (II) oxide. In Fe_2O_3 the iron must have a +3 charge (and a total of +6) to match the -6 charge from the three O^{2-} ions. Iron (III) oxide.
e) The iodide ion has a -1 charge. In AuI the gold must have a +1 charge to match the -1 charge of the one I⁻ ion. Gold (I) iodide. In AuI_3, the gold must have a +3 charge to match the -3 total charge of the three I⁻ ions. Gold (III) iodide.

4.11   In each case, the *ous* ending will be added to the stem of the metal name to indicate the lower charge. The ending *ic* will be added to the root of the metal name to indicate the higher charge. The actual charges on the metals makes no difference, the endings simply indicate the lower and higher charges of the same metal in two different compounds.
a) In $CrCl_2$ the Cr has a +2 charge, and in $CrCl_3$ the Cr has a +3 charge. Thus, $CrCl_2$ is named chromous chloride, and $CrCl_3$ is named chromic chloride.
c) In FeO the iron has a +2 charge, and in Fe_2O_3 iron has a +3 charge, Thus, FeO is named ferrous oxide, and Fe_2O_3 is named ferric oxide.
e) In AuI the gold has a +1 charge, and in AuI_3 the gold has a +3 charge. Thus, AuI is named aurous iodide, and AuI_3 is name auric iodide.

4.12   a) The manganese has a +2 charge as indicated by the (II), and the chloride ion has a -1 charge. Thus, the Mn and Cl must be present in a 1:2 ratio respectively. $MnCl_2$ is the formula.
c) The chromium has a +2 charge as indicated by the (II), and the oxide ion has a -2 charge. Thus, the Cr and O must be present in a 1:1 ratio. CrO is the formula.
e) The tin has a +2 charge as indicated by the (II), and the chloride ion has a -1 charge. Thus, the Sn and Cl must be present in a 1:2 ratio respectively. $SnCl_2$ is the formula.
g) The lead has a +2 charge as indicated by the (II), and the oxide ion has a -2 charge. Thus, the Pb and O must be present in a 1:1 ratio. PbO is the formula.
i) The copper has a +1 charge as indicated by the (I), and the nitride ion has a -3 charge. Thus, the Cu and N must be present in a 3:1 ratio respectively. Cu_3N is the formula.

4.13 a) Each I atom has seven electrons in the valence shell. Thus, each atom shares one of its valence-shell electrons to form a pair that serves both atoms and bonds them together to form a molecule. See Appendix B for the representation.

c) Each At atom has seven electrons in the valence shell. Thus, each atom shares one of its valence-shell electrons to form a pair that serves both atoms and bonds them together to form a molecule. See Appendix B for the representation.

4.14 b) Both I and Br atoms have seven valence-shell electrons. Each atom will share one of its valence-shell electrons to form a shared pair that serves both atoms and bonds them together to form a molecule. See Appendix B for the representation.

d) Each S atom has six valence-shell electrons, and each H atom has one. Each H needs two electrons to complete the valence shell, and so each H shares its electron with one electron from S. In this way the S achieves a filled valence shell of eight electrons, and each H achieves a filled valence shell of two. See Appendix B for the representation.

f) Each H atom has one valence-shell electron, and each Br has seven. One electron from an H atom is shared with one from a Br atom to allow each atom to complete a valence shell (two electrons for H and eight electrons for Br) and bond the atoms together. See Appendix B of the text for the representation.

4.15 a) The C atom has four valence-shell electrons, and each of the four Cl atoms has seven. This gives a total of 32 electrons that are available to form bonds. Each Cl is bonded to the C with one pair of electrons. This requires eight electrons and leaves 24 to complete the octets on the Cl atoms. Each Cl needs an additional 6 electrons to complete the octet. Thus, the 24 remaining electrons will just complete the octets of the Cl atoms. See Appendix B of the text for the resulting structure.

c) The S atom has six valence-shell electrons, each of the three O atoms also has six, and both H atoms have one. This gives a total of 26 electrons available to bond and satisfy the octets of the respective atoms. Each O atom is bonded to the S atom by one pair, and each H atom is bonded to an O atom by a pair. This requires five pairs or 10 of the available 26 electrons. The remaining 16 electrons is just enough to satisfy the octets of the S atom (it needs one pair), and the three O atoms (one O needs three pair, and two need two pair each). The H atoms valence shell was filled with the necessary one pair when the H atoms bonded with the O atoms. See Appendix B of the text for the resulting structure.

e) Each O atom has six valence-shell electrons, which provides a total of 18 electrons to form the bonds. The three O atoms are bonded into a chain using two pair of electrons. This leaves seven pair to satisfy the octets of the three O atoms. Two of the atoms (the atoms on the end of the chain) need three pair each, and the central atom needs two pair. Thus, eight pair are needed, but only seven pair are available. This problem is overcome by sharing two pair between two of the O atoms. See Appendix B of the text for the resulting structure.

Forces Between Particles 51

4.16 The answers given in Appendix B of the text contain the explanations.

4.17 a) Cl and I are in the same group of the periodic table, but Cl is higher in the group and thus has a greater electronegativity.
c) Sr is a metal located on the left in the periodic table, while S is a nonmetal located much farther to the right in the table than Sr. Thus, S is more electronegative.
e) Both Bi and C are located about the same distance to the right in the periodic table, but Bi is lower in the table than is C. Thus, C should have the greater electronegativity.

4.18 Bonds will be polarized when they are located between atoms that have different electronegativities. In general, this means that bonds between non-identical atoms will be polarized.
a) H and I have different electronegativities, so the bond will be polarized. The I, with the higher electronegativity will have a partial negative charge. See Appendix B of the text for the structure.
c) Because the atoms are all identical, no bond polarization takes place.
e) H and Se have different electronegativities, so both bonds will be polarized. The Se is more electronegative than H, and will have a partial negative charge. See Appendix B of the text for the structure.

4.19 In each case, the difference in electronegativity between the bonded atoms is calculated. The compound is then classified according to the value obtained and the categories given by Figure 4.7 of the text.

4.20 In each case, the symmetry of the charge distribution that results in molecules that contain polarized bonds must be noted. If the resulting charge distribution is symmetrical in space, the molecule is nonpolar. If the charge distribution is nonsymmetrical, the molecule is polar. Only molecules that contain polarized bonds can be polar. Molecules that contain no polarized bonds must be nonpolar. See Appendix B of the text for the structures.

4.21 In each case the difference in electronegativity between the bonded atoms is obtained by reference to Table 4.3. The resulting value for the difference (ΔEN) is then used to classify the bonds according to the categories given by Figure 4.7 of the text.

4.22 In each case the more electronegative element of a bonded pair will acquire a partial negative charge. The symmetry of the resulting charge distributions within the molecules determine the polarity of the molecules.
a) O is more electronegative than C, so O has a partial negative charge, and C has a partial positive charge. The charge distribution is not symmetrical, so the molecule is polar.

52   CHAPTER 4

c) I is more electronegative than Al, so the I atoms will have a partial negative charge, and the Al atom will have a partial positive charge. The negative charges are symmetrically distributed about the positive charge, so the molecule is nonpolar.

e) F is more electronegative than O, so the F atoms will have a partial negative charge, and the O atom will have a partial positive charge. The negative charges are symmetrically distributed on opposite sides of the O atom, so the molecule is nonpolar.

4.23 In each case the name is assigned by naming the first-listed element (the one with lowest electronegativity) and following it with the stem + *ide* for the second-listed element (the one of greatest electronegativity). The number of atoms of each element is indicated by using Greek prefixes. The only exception to the use of the Greek prefixes is the prefix mono is generally not included when it would appear at the beginning of the name.
a) N_2O is named dinitrogen monoxide
c) N_2O_4 is named dinitrogen tetroxide (the a in the prefix tetra was dropped for phonetic reasons).
e) PCl_3 is named phosphorus trichloride
g) SiF_4 is named silicon tetraflouride
i) AlI_3 is named aluminum triiodide

4.24 b) The C atom has four valence-shell electrons, and the N atom has five. The negative charge on the ion indicates that one additional electron is available for use in bonding. Thus, the total number of electrons available is 10 or 5 pair. The C and N are bonded together by sharing 3 pair, and one of the remaining 2 pair is put onto the C atom and one onto the N atom to complete their valence shells. See Appendix B of the text for the structure.

d) The P atom has five valence-shell electrons, and each H atom has one. The +1 charge on the ion indicates that one less electron than the total in the valence shells of the bonded atoms is available for bonding. So, a total of 8 electrons is available. One pair is used to bond each of the four H atoms to the P atom. This completes the four pair needed by the P atom to complete its octet, and the one pair needed by each H atom to complete their valence shells. See Appendix B of the text for the structure.

f) The H atom has one valence-shell electron, the S atom has six as does each of the four O atoms. This gives a total of 31 electrons available for bonding. However, the -1 charge on the ion indicates that one additional electron is also available to give a total of 32. Four pair (8 electrons) are used to form a bond between the S atom and each of the four O atoms. An additional pair is used to form a bond between the H atom and one of the O atoms. This leaves 22 electrons to satisfy the valence shells of the O atoms (the valence shell of the S atom was satisfied when bonds were formed with the four O atoms, and the valence shell of the H

Forces Between Particles   53

atom is satisfied with the pair used to bond the H to one of the O atoms). Three of the O atoms need three pair, and one of them (the one bonded to H) needs two pair. Thus, 11 pair or 22 electrons will be needed; exactly the number that remain to be used. See Appendix B of the text for the structure.

4.25   In each case, refer to the structure given in Appendix B of the text as an answer to exercise 4.24.
b) The ion contains only two atoms and so must be linear.
d) The P atom has four pair attached. These four pair will take the shape of a tetrahedron according to VSEPR theory, so the H atoms will form a tetrahedron around the P atom. The geometry will be tetrahedral.
f) The S atom has four pair attached. These four pair will take the shape of a tetrahedron according to VSEPR theory, so the O atoms will form a tetrahedron around the S atom. The geometry will be tetrahedral with the H atom attached to one of the O atoms that form the tetrahedron around the S atom.

4.26   In each case, refer to Table 4.6 for the names and charges of the common polyatomic ions. The name in each case is the name of the metal followed by the name of the polyatomic ion.
a) Potassium, K, is a group IA(1) element and so forms a K^+ ion. The cyanide ion, CN^-, has a -1 charge, so the two combine in a 1:1 ratio. The formula is KCN, and the name is potassium cyanide.
c) Calcium, Ca, is a group IIA(2) metal and so forms a Ca^{2+} ion. The phosphate ion, PO_4^{3-} has a -3 charge. Thus, the two will combine in a 3:2 ratio respectively to balance the charges of the two ions. The formula is $Ca_3(PO_4)_2$, and the name is calcium phosphate.
e) Barium, Ba, is a group IIA(2) metal and so forms a Ba^{2+} ion. The sulfate ion, SO_4^{2-}, has a -2 charge. Thus, the two ions will combine in a 1:1 ratio. The formula is $BaSO_4$ and the name is barium sulfate.
g) Lithium, Li, is a group IA(1) metal and so forms an Li^+ ion. The carbonate ion, CO_3^{2-}, has a -2 charge. Thus, the two ions will combine in a 2:1 ratio respectively to balance the charges of the two ions. The formula is Li_2CO_3 and the name is lithium carbonate.
i) Beryllium, Be, is a group IIA(2) metal and so forms a Be^{2+} ion. The phosphate ion, PO_4^{3-}, has a -3 charge. Thus, the two ions will combine in a 3:2 ratio respectively to balance the charges of the two ions. The formula is $Be_3(PO_4)_2$, and the name is beryllium phosphate.

4.27   In each case the formula and charge of the polyatomic ion are obtained from Table 4.6 of the text.
a) Potassium, K, is a group IA(1) metal and forms a K^+ ion. The carbonate ion, CO_3^{2-}, has a -2 charge. Thus, the two ions will combine in a 2:1 ratio respectively to balance the ionic charges. The formula is K_2CO_3.

54   CHAPTER 4

   c) Calcium, Ca, is a group IIA(2) metal and forms a Ca^{2+} ion. The acetate ion, $C_2H_3O_2^-$, has a -1 charge. Thus, the two ions will combine in a 1:2 ratio respectively to balance the ionic charges. The formula is $Ca(C_2H_3O_2)_2$.
   e) Sodium, Na, is a group IA(1) metal and forms an Na^+ ion. The bisulfate ion, HSO_4^-, has a -1 charge. Thus, the two ions will combine in a 1:1 ratio to balance the ionic charges. The formula is $NaHSO_4$.
   g) Calcium, Ca, is a group IIA(2) metal and forms a Ca^{2+} ion. The hydroxide ion, OH^-, has a -1 charge. Thus, the two ions will combine in a 1:2 ratio respectively to balance the ionic charges. The formula is $Ca(OH)_2$.
   i) The ammonium ion is a polyatomic positive ion with the formula NH_4^+. The dihydrogenphosphate ion, $H_2PO_4^-$, has a -1 charge. Thus, the two ions will combine in a 1:1 ratio to balance the ionic charges. The formula is $NH_4(H_2PO_4)$.

4.28 b) A group IA(1) metal will form an M^+ ion and will combine with the $C_2H_3O_2^-$ ion in a 1:1 ratio respectively to balance the ionic charges. The formula is $MC_2H_3O_2$.
   d) A group IIA(2) metal will form an M^{2+} ion and will combine with the HPO_4^{2-} ion in a 1:1 ratio respectively to balance the ionic charges. The formula is $MHPO_4$.
   f) An M^{2+} ion will combine with a $Cr_2O_7^{2-}$ ion in a 1:1 ratio to balance the ionic charges. The formula is MCr_2O_7.
   h) An M^{3+} ion will combine with an NO_3^- ion in a 1:3 ratio respectively to balance ionic charges. The formula is $M(NO_3)_3$.

4.29 The explanation is given in Appendix B of the text.

4.31 The explanation is given in Appendix B of the text.

4.33 The explanation is given in Appendix B of the text.

SELF-TEST QUESTIONS

Multiple Choice

1.   Which of the following elements has the lowest electronegativity?
     a) As   b) P   c) Br   d) Cl

2. In describing the strength of interparticle forces we discover that the weakest forces or bonds are
   a) covalent
   b) metallic
   c) ionic
   d) dipolar
   e) dispersion

3. The formula for the compound formed between the elements Ba and O would be
   a) BaO
   b) Ba$_2$O
   c) BaO$_2$
   d) Ba$_2$O$_3$

4. The formula for the ionic compound containing Al^{3+} and SO$_4^{2-}$ ions would be
   a) Al(SO$_4$)$_2$
   b) AlSO$_4$
   c) Al$_3$(SO$_4$)$_2$
   d) Al$_2$(SO$_4$)$_3$

5. The expected formula of the molecule formed when nonmetals C and H combine in compliance with the octet rule is
   a) CH$_4$
   b) CH$_2$
   c) C$_4$H
   d) CH$_3$

6. The name of the covalent compound PCl$_3$ is
   a) trichlorophosphide
   b) phosphorus trichloride
   c) phosphorus trichlorine
   d) phosphorus chloride

7. The compound MgSO$_4$ is correctly named
   a) magnesium sulfur tetroxide
   b) magnesium sulfoxide
   c) magnesium sulfide
   d) magnesium sulfate

8. This bond is found in molecules such as HCl and H$_2$O.
   a) nonpolar covalent bond
   b) polar covalent bond
   c) ionic bond
   d) metallic bond

9. If the electronegativity difference between two elements A and B is 1.0, what type of bond is A-B?
   a) nonpolar covalent
   b) polar covalent
   c) ionic
   d) metallic

10. Which of the following is a correct electron dot formula for sulfur (element 16)?
    a) :S̈:
    b) S̈
    c) ·S̈·
    d) ·S̈:

56 CHAPTER 4

11. In the structures below, only bonding electrons are shown, and they are denoted by a dash. A correct structure of SO_3^{2-} is

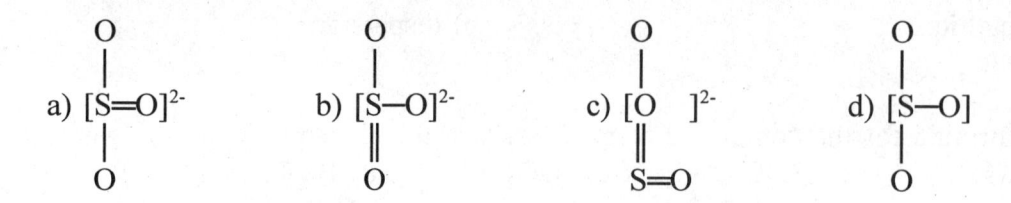

a) $\begin{array}{c} O \\ | \\ [S{=}O]^{2-} \\ | \\ O \end{array}$
b) $\begin{array}{c} O \\ | \\ [S{-}O]^{2-} \\ \| \\ O \end{array}$
c) $\begin{array}{c} O \\ | \\ [O \quad]^{2-} \\ \| \\ S{=}O \end{array}$
d) $\begin{array}{c} O \\ | \\ [S{-}O] \\ | \\ O \end{array}$

12. A covalent molecule forms between elements A and B. B is more electronegative. Which of the following molecules would be polar?

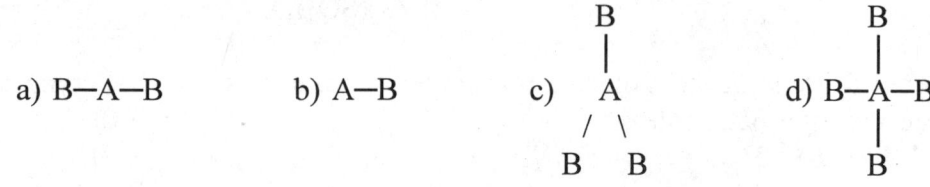

a) B—A—B
b) A—B
c) $\begin{array}{c} B \\ | \\ A \\ / \ \backslash \\ B \quad B \end{array}$
d) $\begin{array}{c} B \\ | \\ B{-}A{-}B \\ | \\ B \end{array}$

13. Which of the following molecules contains all polarized bonds, but is nonpolar?
    a) $\begin{array}{c} H{-}S \\ \quad \backslash \\ \quad H \end{array}$
    b) O=C=O
    c) H—Cl
    d) F—F

True-False

14. No more than one pair of electrons can be shared to form covalent bonds between atoms.

15. Dispersion forces between particles are correctly classified as very strong.

16. A compound between elements with atomic numbers 7 and 8 will contain covalent bonds.

17. All covalent bonds are polar.

18. The interparticle forces in a solid noble gas would have to be polar in nature.

19. Argon (Ar) has a higher boiling point than Kr.

Matching

An ionic compound is formed from each of the pairs given on the left. For each pair, choose the correct formula for the resulting compound from the responses on the right.

20. X^+ and Y^{2-}

21. X^+ and Y^-

22. X^{3+} and Y^{3-}

23. X is a group IIA (2) ion and Y is a group VIA (16) ion

a) X_3Y_2
b) XY
c) XY_2
d) X_2Y

For each molecule given on the left, predict the molecular geometry based on VSEPR theory.

24. NH_3

25. BrCl

26. H_2S

27. CO_2

28. CH_2

a) linear
b) planar triangle
c) triangular pyramid
d) tetrahedral
e) bent

For each of the molecules on the left, choose the statement from the right that correctly gives the polarity of the bonds in the molecule and the polarity of the molecule as a whole.

29. H_2S, H—S
           \
            H

30. CO_2, O=C=O

31. N_2O, N≡N—O

32. O_3,   O
            / \
           O   O

a) a polar molecule containing all polarized bonds
b) a nonpolar molecule containing all nonpolarized bonds
c) a nonpolar molecule containing all polarized bonds
d) a polar molecule containing polarized bonds and nonpolarized bonds

ANSWERS TO PROGRAMMED REVIEW

4.1 a) stable   b) two   c) eight

4.2 a) octet rule   b) simple ions   c) ionic bond   d) lose
     e) gain   f) isoelectronic

4.3 a) binary   b) combining ratio   c) crystal   d) lattice

4.4 a) sharing   b) covalent bond

4.5 a) three-dimensional   b) double and triple   c) VSEPR
     d) central

4.6 a) nonpolar   b) electronegativity   c) polar   d) polar molecules

4.7 a) polyatomic   b) negatively

4.8 a) network   b) metallic   c) kernels   d) dipolar
     e) hydrogen bonding   f) dispersion

ANSWERS TO SELF-TEST QUESTIONS

1.	a	12.	b	23.	b
2.	e	13.	b	24.	c
3.	a	14.	F	25.	a
4.	d	15.	F	26.	e
5.	a	16.	T	27.	a
6.	b	17.	F	28.	d
7.	d	18.	F	29.	a
8.	b	19.	F	30.	c
9.	b	20.	d	31.	d
10.	d	21.	b	32.	b
11.	d	22.	b		

CHAPTER 5

Chemical Reactions

PROGRAMMED REVIEW

Section 5.1 Chemical Equations

According to convention, a chemical equation is written with (a) _____ on the left and (b) _____ on the right. In a (c) _____ equation the total number of each kind of atom is equal in the reactants and products.

Section 5.2 Types of Reactions

In this text reactions are first classified as (a) _____ or (b) _____. Both of these types are further classified as (c) _____ or (d) _____. Only redox reactions are further classified as (e) _____ _____ reactions, and only nonredox are further classified as (f) _____ _____ reactions.

Section 5.3 Redox Reactions

One meaning of the term (a) _____ is to lose electrons. The term (b) _____ can mean to combine with hydrogen. (c) _____ _____ are positive or negative numbers assigned to the elements in chemical formulas according to a set of rules. The oxidation number of an uncombined element is always (d) _____. In a redox reaction, the substance oxidized is called the (e) _____ agent, and the substance reduced is called the (f) _____ agent.

Section 5.4 Decomposition Reactions

In decomposition reactions a single substance forms two or more (a) _____ substances. Decomposition reactions can be either (b) _____ or (c) _____.

Section 5.5 Combination Reactions

Combination reactions are also called (a) _____ or (b) _____ reactions. In combination reactions, (c) _____ or _____ substances combine to form a (d) _____ substance.

Section 5.6 Replacement Reactions

(a) _____ replacement or (b) _____ reactions are always redox reactions, but (c) _____ replacement or (d) _____ reactions are not redox. Partner swapping is a characteristic of (e) _____ replacement reactions.

Section 5.7 Ionic Equations

When ionic substances dissolve in water they (a) _____ _____ into their constituent (b) _____. (c) _____ equations contain no ions. In a (d) _____ equation all substances that form ions are written as ions. Ions that appear on both the reactant and product sides of equations are called (e) _____ _____, and are not shown in (f) _____ _____ equations.

Section 5.8 Energy and Reactions

In addition to composition changes, (a) _____ changes accompany all chemical reactions. The energy of reactions can take the form of heat, and when heat is liberated the reaction is called (b) _____. When heat is absorbed the reaction is called (c) _____.

Section 5.9 The Mole and Chemical Equations

Application of the (a) _____ concept to a balanced equation provides several (b) _____ from which (c) _____ can be obtained that are useful in solving problems using the (d) _____ _____ method.

Section 5.10 The Limiting Reactant

According to the (a) _____ _____ principle, the maximum amount of product that can be obtained from a mixture of reactants is determined by the amount of (b) _____ reactant present. As soon as the (c) _____ reactant has all reacted, the reaction will (d) _____.

Section 5.11 Reaction Yields

Reactions that do not give the desired products are called (a) _____ reactions. Such reactions are one reason that the (b) _____ yield of a reaction is often less than the (c) _____ yield. The (d) _____ yield of a reaction is the actual yield divided by the theoretical yield and multiplied by 100.

Chemical Reactions 61

SOLUTIONS TO EXERCISES ANSWERED IN THE TEXT

5.1 Reactants are the materials located to the left of the arrow, and products are located to the right.
   b) Reactant is H_2O_2; the products are H_2O and O_2
   d) Reactants are copper(II) oxide and hydrogen; the products are copper and water.
   f) The reactant is $KClO_3$; the products are KCl and O_2

5.2 The number of atoms on each side of the arrow must be the same to be consistent with the law of conservation of matter.
   b) P_4 on the left indicates the presence of four phosphorus atoms.
   O_2 on the left indicates the presence of two oxygen atoms.
   The formula P_4O_{10} indicates the presence of four phosphorus and ten oxygen atoms on the right. Not consistent.
   d) The formula CH_4 indicates the presence of one carbon atom and four hydrogen atoms on the left.
   The $2O_2$ indicates the presence of four oxygen atoms on the left.
   The formula CO_2 indicates the presence of one carbon atom and two oxygen atoms on the right.
   The $2H_2O$ indicates the presence of four hydrogen atoms and two oxygen atoms on the right. The total number of each kind of atom is seen to be the same on the right and the left. Consistent.
   f) The Cl_2 on the left indicates the presence of two chlorine atoms.
   The I^- on the left indicates the presence of one iodine atom in the form of an ion.
   The I_2 on the right indicates the presence of two iodine atoms.
   The $2Cl^-$ on the right indicates the presence of two chlorine atoms in the form of ions. Not consistent.

5.3 A balanced equation is one that is consistent with the law of conservation of mass. That is, the number of each kind of atom must be the same on both sides of the equation.
   a) The formula H_2S on the left indicates the presence of two hydrogen atoms and one sulfur atom. The I_2 on the left indicates the presence of two iodine atoms. The $2HI$ on the right indicates the presence of two hydrogen atoms and two iodine atoms. The S on the right indicates the presence of one sulfur atom. The number of each kind of atom is the same on each side of the equation, so the equation is balanced.
   c) The formula SO_2 on the left indicates the presence of one sulfur atom and two oxygen atoms. The formula H_2O on the left indicates the presence of two hydrogen atoms and one oxygen atom. The formula H_2SO_3 on the right indicates the presence of two hydrogen atoms, one sulfur atom, and three oxygen atoms. Thus, the number of each kind of atom is the same on each side of the equation, and the equation is balanced.

62   CHAPTER 5

e) The Cu on the left indicates the presence of one copper atom. The formula H_2SO_4 on the left indicates the presence of two hydrogen atoms, one sulfur atom and four oxygen atoms. The formula $CuSO_4$ on the right indicates the presence of one copper atom, one sulfur atom and four oxygen atoms. The formula SO_2 on the right indicates the presence of one sulfur atom and two oxygen atoms. The formula H_2O on the right indicates the presence of two hydrogen atoms and one oxygen atom. In this case, both the number of sulfur atoms and the number of oxygen atoms are seen to be different on the two sides of the equation. The equation is not balanced.

5.4   Equations are balanced by adjusting the number of atoms of each kind so they are the same on both sides of the equation. However, it must be remembered that the subscripts used to write the formulas of compounds or the formulas of the molecules of elements cannot be changed. They are fixed by the combining ratios of the atoms. The coefficients preceding formulas are changed to adjust the numbers of the atoms on the sides of the equation. This is essentially a trial and error approach, but it must be remembered to check the final coefficients of an equation balanced this way to make certain that the lowest coefficients possible are used. See Appendix B of the text for the balanced equations.

5.5   Various rules are used to assign or calculate the oxidation number of elements. The rules are given in the text.

b) The oxidation number of F in F_2 is 0 because F_2 is an element and rule 1 says the oxidation number of uncombined elements is zero.

d) The P in $H_4P_2O_7$ has an oxidation number of +5. This was calculated as follows: The O.N. of H is +1 (rule 4), and the O.N. of O is -2 (rule 5). According to rule 6, 4(O.N. H) + 2(O.N. P) + 7(O.N. O) = 0. Thus, 4(+1) + 2(O.N. P) + 7(-2) = 0, or (+4)+(-14)+ 2(O.N. P) = 0. This leads to -10 + 2(O.N. P) = 0 or O.N. P = +5

f) The S in H_2S has an oxidation number of -2. This was calculated as follows: The O.N. of H is +1 (rule 4). According to rule 6, 2(O.N. H) + O.N. S = 0, or 2(+1) + O.N. S = 0. This immediately leads to the conclusion that the O.N. of S is -2.

h) The N in N_2O has an oxidation number of +1. This was calculated as follows: The O.N. of O is -2 (rule 5). According to rule 6, 2(O.N. N) + O.N. O = 0, or 2(O.N. N) + (-2) = 0. This leads to the conclusion that the oxidation number of N is +1.

5.6   In each case, the oxidation number of each element was calculated using rule 6, or assigned using one of the other rules.

b) O.N. of Na = +1 (rule 3), O.N. of N = +5 (rule 6), O.N. of O = -2 (rule 5). The highest is seen to be N (+5).

d) O.N. of H = +1 (rule 4), O.N. of Cl = +7 (rule 6), O.N. of O = -2 (rule 5). The highest is seen to be Cl (+7).
f) O.N. of Ca = +2 (rule 3), O.N. of N = +5 (rule 6), O.N. of O = -2 (rule 5). The highest is seen to be N (+5).
h) O.N. of C = -2 (rule 6), O.N. of H = +1 (rule 4), O.N. of O = -2 (rule 5). The highest is seen to be H (+1).

5.7 In each case, the oxidation number of the element in question is determined on the left side of the equation and on the right side. If the oxidation number increases from the left to the right side, the element has been oxidized. If the oxidation number decreases, the element has been reduced. If no change in oxidation number takes place, the element has been neither oxidized nor reduced.
a) O.N. of Mg on the left is 0, and on the right it is +2. The Mg has been oxidized.
c) O.N. of Ag on the left is +1, and on the right it is +1. The Ag has been neither oxidized nor reduced.
e) O.N. of Zn on the left is 0, and on the right it is +2. The Zn has been oxidized.
g) O.N. of S on the left is +4, and on the right it is +4. The S has been neither oxidized nor reduced.
i) O.N. of C on the left is +2, and on the right it is +4. The C has been oxidized.

5.8 In each case the oxidizing agent is the molecule, ion or atom that contains the element that undergoes reduction (O.N. decreases) as the reaction takes place from left to right. The reducing agent is the molecule, ion or atom that contains the element that undergoes oxidation (O.N. increases) as the reaction takes place from left to right. The answers in appendix B of the text show all the oxidation numbers and identify the oxidizing and reducing agents along with the element that changed in oxidation number.

5.9 The oxidizing agent is the molecule or atom that contains the element whose oxidation number decreases as the reaction takes place from left to right. The reducing agent is the molecule or atom whose oxidation number increases as the reaction takes place from left to right. The answer in Appendix B of the text shows all the oxidation numbers and identifies the oxidizing and reducing agents.

5.11 The oxidizing agent is the molecule that contains the element that is reduced as the reaction goes from left to right. The reducing agent is the molecule that contains the element that is oxidized as the reaction goes from left to right. The answer in appendix B of the text shows all the oxidation numbers and identifies the oxidizing and reducing agents.

5.12 In each case, the reaction will be classified as a redox if elements change in oxidation number as the reaction goes from the left to the right. Each reaction is classified further using the patterns shown in equations 5.13, 5.16, 5.19 and 5.21 of the text.

64   CHAPTER 5

> b) The oxidation number of Ag changes from +1 to 0, while the oxidation number of oxygen changes from -2 to 0 as the reaction takes place. The reaction is a redox reaction. In the reaction, one material changes into two new materials, matching the pattern of equation 5.13. Thus, the reaction is a decomposition reaction.
>
> d) The oxidation number of Na changes from 0 to +1, while the oxidation number of oxygen changes from 0 to -2 as the reaction takes place. The reaction is a redox reaction. In the reaction, two materials change into a single material, matching the pattern of equation 5.16. Thus, the reaction is a combination reaction.
>
> f) The oxidation number of Zr changes from +4 to 0, while the oxidation number of I changes from -1 to 0 as the reaction takes place. The reaction is a redox reaction. In the reaction, one material changes into two new materials, matching the pattern of equation 5.13. Thus, the reaction is a decomposition reaction.
>
> h) The oxidation number of Cr changes from +3 to 0, while the oxidation number of Al changes from 0 to +3 as the reaction takes place. The reaction is a redox reaction. In the reaction, one element, Al, displaces another element, Cr, from a compound and takes its place. This matches the pattern of equation 5.19 and corresponds to a single-replacement reaction.
>
> j) No oxidation numbers change as this reaction takes place. Thus, it is a non-redox reaction. In the reaction, one material changes into two new materials, matching the pattern of equation 5.13. The reaction is a decomposition reaction.
>
> l) The oxidation number of Zn changes from 0 to +2, while the oxidation number of Cl changes from 0 to -1 as the reaction takes place. The reaction is a redox reaction. In the reaction, two elements combine to form a compound, matching the pattern of equation 5.16. The reaction is a combination reaction.

5.14   No oxidation numbers change as the reaction takes place. The O.N. of Na is +1 on the left and right, the O.N. of H is +1 on the left and the right, the O.N. of C is +4 on the left and the right, and the O.N. of O is -2 on the left and the right. Thus, the reaction is nonredox.

5.16   The oxidation number of H in H_2O_2 is +1, while the oxidation number of O is -1. Note that this oxidation number of oxygen is one of the exceptions given in rule 5. The oxidation number of H in H_2O is +1; the oxidation number of O in H_2O is -2, and the oxidation number of O in O_2 is 0. Thus, we see that the oxygen has been both oxidized and reduced as the reaction took place. The reaction is a redox reaction.

5.18   No oxidation numbers change as the reaction takes place. The oxidation numbers of the elements on both the left and right side of the equation are: Ca = +2, P = +5, O = -2, and H = +1. The reaction is nonredox. The reaction matches the pattern of equation 5.16 and is a combination reaction.

Chemical Reactions 65

5.19 Table 4.6 give the formulas and charges of some common polyatomic ions that are found in the ionic substances in the exercise. It must be remembered that the polyatomic ions will not split into parts but remain intact. Appendix B of the text lists the ions that will form in each case.

5.20 In each case, all substances that dissolve are assumed to form ions when they dissolve. Any ions that appear on both sides of the equation are classified as spectator ions and will not appear in the net ionic equation. The answers given in Appendix B of the text show the ionic equations, identify the spectator ions and show the net ionic equations.

5.21 In each case, the ionic equation, identity of spectator ions and net ionic equation are given in Appendix B of the text.

5.22 The explanation is given as the answer in Appendix B of the text.

5.24 The explanation is given as the answer in Appendix B of the text.

5.26 The statements are written as the answers in Appendix B of the text.

5.27 The statements are written as the answers in Appendix B of the text.

5.29 The known or given quantity is 500 g CaO, and the unit of the unknown quantity is g $CaCO_3$. The factor that will cancel units properly is 100.1 g $CaCO_3$/56.1 g CaO. The factor came from the statement: 100.1 g $CaCO_3$ → 56.1 g CaO + 44.0 g CO_2.

The solution is:

$$500 \text{ g CaO} \times \frac{100.1 \text{ g } CaCO_3}{56.1 \text{ g CaO}} = 892 \text{ g } CaCO_3$$

The answer has been rounded to the correct number of significant figures.

5.31 The known or given quantity is 96.0 g CaC_2, and the unit of the unknown quantity is g H_2O. The factor that will cancel units properly is 36.0 g H_2O/64.1 g CaC_2. The factor came from the statement:

64.1 g CaC_2 + 36.0 g H_2O → 74.1 g $Ca(OH)_2$ + 26.0 g C_2H_2.

The solution is:

66 CHAPTER 5

$$96.0 \text{ g CaC}_2 \times \frac{36.0 \text{ g H}_2O}{64.1 \text{ g CaC}_2} = 53.9 \text{ g H}_2O$$

The answer has been rounded to the correct number of significant figures.

5.33 The known or given quantity is 96.0 g CaC_2, and the unit of the unknown quantity is mol C_2H_2. The factor that will cancel units properly is 1 mol C_2H_2/64.1 g CaC_2. The factor came from two statements:

$$64.1 \text{ g CaC}_2 + 36.0 \text{ g H}_2O \rightarrow 74.1 \text{ g Ca(OH)}_2 + 26.0 \text{ g C}_2H_2 \quad \text{and}$$
$$1 \text{ mol CaC}_2 + 2 \text{ mol H}_2O \rightarrow 1 \text{ mole Ca(OH)}_2 + 1 \text{ mole C}_2H_2$$

The solution is:

$$95.0 \text{ g CaC}_2 \times \frac{1 \text{ mol C}_2H_2}{64.1 \text{ g CaC}_2} = 1.50 \text{ mol C}_2H_2$$

The answer has been rounded to the correct number of significant figures.

5.35 The known or given quantity is 1.00 kg Ti, and the unit of the unknown quantity is g Mg. The factor that will cancel units properly is 48.6 g Mg/47.9 g Ti. When using this factor, it will be necessary to express the known quantity as 1.00×10^3 g rather than as 1.00 kg. The factor came from the statement:

$$189.7 \text{ g TiCl}_4 + 48.6 \text{ g Mg} \rightarrow 47.9 \text{ g Ti} + 190.4 \text{ g MgCl}_2$$

The solution is:

$$1.00 \times 10^3 \text{ g Ti} \times \frac{48.6 \text{ g Mg}}{47.9 \text{ g Ti}} = 1.02 \times 10^3 \text{ g Mg}$$

The answer has been rounded to the correct number of significant figures.

5.37 The known or given quantity is 1.00 mol $C_6H_{12}O_2$, and the unit of the unknown quantity is g O_2. The factor that will cancel units properly is 256.0 g O_2/1 mol $C_6H_{12}O_2$. The factor came from two statements:

$$116.2 \text{ g C}_6H_{12}O_2 + 256.0 \text{ g O}_2 \rightarrow 264.1 \text{ g CO}_2 + 108.1 \text{ g H}_2O \quad \text{and}$$

$$1 \text{ mol C}_6H_{12}O_2 + 8 \text{ mol O}_2 \rightarrow 6 \text{ mol CO}_2 + 6 \text{ mol H}_2O$$

The solution is:

Chemical Reactions 67

$$1.00 \text{ mol } C_6H_{12}O_2 \times \frac{256.0 \text{ g } O_2}{1 \text{ mol } C_6H_{12}O_2} = 256 \text{ g } O_2$$

The answer has been rounded to the correct number of significant figures.

5.39 The first step in solving the problem is to convert both given or known quantities to the same units. The 320.0 g of Br_2 is converted to mol Br_2 by the following calculation, where the atomic weight of Br was obtained from the periodic table.

$$320.0 \text{ g } Br_2 \times \frac{1 \text{ mol } Br_2}{159.8 \text{ g } Br_2} = 2.00 \text{ mol } Br_2$$

The problem is now solved by calculating the number of moles of HBr that could be produced from each of the reactants assuming the other reactant is present in excess. The limiting reactant will be the one that will produce the least amount of product.

$$1.50 \text{ mol } H_2 \times \frac{2 \text{ mol } HBr}{1 \text{ mol } H_2} = 3.00 \text{ mol } HBr$$

$$2.00 \text{ mol } Br_2 \times \frac{2 \text{ mol } HBr}{1 \text{ mol } Br_2} = 4.00 \text{ mol } HBr$$

These results indicate that the limiting reactant is the hydrogen, H_2. The calculation shows that only 3.00 moles of the HBr product could be formed. This quantity is converted to grams by the following calculation:

$$2.00 \text{ mol } HBr \times \frac{80.9 \text{ g } HBr}{1 \text{ mol } HBr} = 243 \text{ g } HBr$$

The answer has been rounded to the correct number of significant figures.

5.41 The first step in solving this problem is to express all the given or known quantities in the same units. We will express them in moles, so 34.0 g NH_3 must be converted to moles.

$$34.0 \text{ g } NH_3 \times \frac{1 \text{ mol } NH_3}{17.0 \text{ g } NH_3} = 2.00 \text{ mol } NH_3$$

The next step is to calculate the amount of product that could be produced from the amount of each of the reactants, assuming the other reactants are present in excess. The reactant that would produce the least amount of product is the limiting reactant.

68 CHAPTER 5

$$2.00 \text{ mol } NH_3 \times \frac{1 \text{ mol } NH_4HCO_3}{1 \text{ mol } NH_3} = 2.00 \text{ mol } NH_4HCO_3$$

$$3.00 \text{ mol } CO_2 \times \frac{1 \text{ mol } NH_4HCO_3}{1 \text{ mol } CO_2} = 3.00 \text{ mol } NH_4HCO_3$$

$$1.50 \text{ mol } H_2O \times \frac{1 \text{ mol } NH_4HCO_3}{1 \text{ mol } H_2O} = 1.50 \text{ mol } NH_4HCO_3$$

The H_2O is seen to be the limiting reactant. The 1.50 moles of NH_4HCO_3 is now converted into grams.

$$1.50 \text{ mol } NH_4HCO_3 \times \frac{79.1 \text{ g } NH_4HCO_3}{1 \text{ mol } NH_4HCO_3} = 119 \text{ g } NH_4HCO_3$$

5.43 The % yield is calculated as follows:

$$\% \text{ yield} = \frac{\text{actual yield}}{\text{theoretical yield}} \times 100$$

$$\% \text{ yield} = \frac{9.14 \text{ g}}{11.98 \text{ g}} \times 100 = 76.3\%$$

5.45 In a combination reaction, the reactants combine to form a single product. According to the information given, 6.36 g of A would exactly react with 3.21 g of B. Thus, the mass of the product would theoretically be the sum of the masses of the two reactants, or 6.36 g + 3.21 g, or 9.57 g. This is the theoretical yield of the reaction. The actual yield was found to be 8.47 g of product. The percentage yield of the reaction is calculated as follows:

$$\% \text{ yield} = \frac{\text{actual yield}}{\text{theoretical yield}} \times 100 = \frac{8.47 \text{ g}}{9.57 \text{ g}} \times 100 = 88.5\%$$

5.47 In order to solve this problem the theoretical yield must first be calculated. This calculation involves using the 7.22 g sample of HgO as the known or given quantity and solving for the mass of mercury, Hg, that could be prepared from it based on the given equation. The factor that will cancel units properly is 401.2 g Hg/433.2 g HgO.

$$7.22 \text{ g HgO} \times \frac{401.2 \text{ g Hg}}{433.2 \text{ g HgO}} = 6.69 \text{ g Hg}$$

Thus, the theoretical yield of the reaction is 6.69 g Hg. The actual yield was 5.95 g Hg. The percentage yield is calculated as follows:

$$\% \text{ yield} = \frac{actual\ yield}{theoretical\ yield} \times 100 = \frac{5.95 \text{ g}}{6.69 \text{ g}} \times 100 = 88.9\%$$

SELF-TEST QUESTIONS

Multiple Choice

1. Which of the following is a reactant in the reaction:

    $$2H^+ + CaCO_3 \rightarrow H_2O + Ca^{2+} + CO_2$$

    a) H_2O b) H^+ c) Ca^{2+} d) CO_2

2. What is the coefficient to the left of H_2 when the following equation is balanced?

    $$Na + H_2O \rightarrow NaOH + H_2$$

    a) 1   b) 2   c) 3   d) 4

3. A decomposition reaction can also be classified as
    a) a combination reaction
    b) a double replacement reaction
    c) a single replacement reaction
    d) a redox or nonredox reaction

4. The oxidation number of a monoatomic ion is always
    a) +1
    b) -1
    c) 0
    d) equal to the charge on the ion

5. In a redox reaction, the reducing agent
   a) is reduced
   b) is oxidized
   c) gains electrons
   d) contains an element whose oxidation number decreases

6. Identify the spectator ion in the following reaction:

   $$Cl_2 + 2K^+ + 2Br^- \rightarrow 2K^+ + 2Cl^- + Br_2$$

   a) Br⁻   b) Cl_2 c) K^+ d) Cl⁻

7. Which of the following statements is consistent with the balanced equation $2SO_2 + O_2 \rightarrow 2SO_3$?

   a) one mol SO_2 reacts with one mol O_2
   b) two mol SO_2 produces two mol SO_3
   c) 64.1 g SO_2 reacts with 32.0 g O_2
   d) 32.0 g O_2 produces 80.1 g SO_3

8. According to the reaction $N_2 + 3H_2 \rightarrow 2NH_3$, how many grams of H_2 are needed to produce 4.0 moles of NH_3?
   a) 3.0   b) 6.0   c) 9.0   d) 12.0

9. For the reaction $2H_2 + O_2 \rightarrow 2H_2O$, how many moles of H_2O could be obtained by reacting 64.0 grams of O_2 and 8.0 grams of H_2?
   a) 72   b) 18   c) 2.0   d) 4.0

10. Nitrogen and hydrogen react as follows to form ammonia: $N_2 + 3H_2 \rightarrow 2NH_3$. In a mixture containing 1.50 g H_2 and 6.00 g N_2, it is true that
    a) H_2 is the limiting reactant
    b) N_2 is the limiting reactant
    c) H_2 is present in excess
    d) an exact reacting ratio of N_2 and H_2 is present

True-False

11. Oxidation numbers never change in combination reactions.

12. Oxidation numbers of some combined elements may be equal to zero.

13. The loss of hydrogen corresponds to an oxidation process.

14. A reaction that releases heat is classified as endothermic.

15. Balanced equations represent statements of the law of conservation of matter.

16. The percentage yield of a reaction cannot exceed 100%.

Matching

Match each of the reactions below with the correct category from the responses.

17. $3Fe + 2O_2 \rightarrow Fe_3O_4$ a) decomposition
                                              b) single replacement
18. $CuO + H_2 \rightarrow Cu + H_2O$ c) double replacement
                                              d) combination
19. $CuCO_3 \rightarrow CuO + CO_2$

20. $KOH + HBr \rightarrow H_2O + KBr$

21. $2Ag_2O \rightarrow 4Ag + O_2$

Match the oxidation number of the underlined element to the correct value given as a response.

22. $H_2\underline{C}O_3$ a) +1
                                    b) +2
23. $\underline{C}O_2$ c) +3
                                    d) +4
24. $\underline{Cs}_2O$

25. $\underline{Cr}_2O_3$

26. $\underline{S}O_3^{2-}$

27. $\underline{Ba}^{2+}$

72 CHAPTER 5

ANSWERS TO PROGRAMMED REVIEW

5.1 a) reactants b) products c) balanced

5.2 a) redox b) nonredox c) combination d) decomposition
     e) single replacement f) double replacement

5.3 a) oxidation b) reduction c) oxidation numbers d) zero
     e) reducing f) oxidizing

5.4 a) simpler b) redox c) nonredox

5.5 a) addition b) synthesis c) two or more d) single

5.6 a) single b) substitution c) double d) metathesis
     e) double

5.7 a) break apart b) ions c) full d) total ionic
     e) spectator ions f) net ionic

5.8 a) energy b) exothermic c) endothermic

5.9 a) mole b) statements c) factors d) factor-unit

5.10 a) limiting reactant b) limiting c) limiting d) stop

5.11 a) side b) actual c) theoretical d) percentage

ANSWERS TO SELF-TEST QUESTIONS

1.	b	10.	b	19.	a
2.	a	11.	F	20.	c
3.	d	12.	T	21.	a
4.	d	13.	T	22.	d
5.	b	14.	F	23.	d
6.	c	15.	T	24.	a
7.	b	16.	T	25.	c
8.	b	17.	d	26.	d
9.	d	18.	b	27.	b

CHAPTER 6

The States of Matter

PROGRAMMED REVIEW

Section 6.1 Observed Properties of Matter

The (a) _____ of a substance is defined as the mass of a sample divided by the volume of the sample. The change in volume resulting from a change in pressure is called the (b) _____ and is quite high for matter in the (c) _____ state. The change in volume resulting from a temperature change is called the (d) _____ _____ of matter.

Section 6.2 The Kinetic Molecular Theory of Matter

According to the kinetic molecular theory, matter is composed of tiny (a) _____ that are in constant (b) _____ and therefore possess (c) _____ energy. The particles also possess (d) _____ energy as a result of attracting or repelling each other. (e) _____ forces are most closely associated with potential energy, while (f) _____ forces are associated with kinetic energy of the particles.

Section 6.3 The Solid State

In the solid state, (a) _____ forces are stronger than (b) _____ forces. Solids generally have a (c) _____ density, (d) _____ shape, (e) _____ compressibility and (f) _____ _____ thermal expansion.

Section 6.4 The Liquid State

In the liquid state (a) _____ forces dominate slightly. Liquids are characterized by a (b) _____ density, (c) _____ shape, (d) _____ compressibility and (e) _____ thermal expansion.

Section 6.5 The Gaseous State

In the gaseous state, (a) _____ forces completely overcome (b) _____ forces. Gases are characterized by a (c) _____ density, (d) _____ shape, (e) _____ compressibility and (f) _____ thermal expansion.

Section 6.6 The Gas Laws

Mathematical expressions that describe the behavior of gases as they are mixed, subjected to pressure changes, etc. are called (a) _____ _____. An important quantity in gas calculations is the (b) _____ which is defined as a force per unit area. Two units of pressure are the standard (c) _____ and the (d) _____ which are, respectively, the pressure needed to support 760 and 1 mm columns of mercury.

Section 6.7 Pressure, Temperature and Volume Relationships

Boyle's Law describes a relationship between the (a) _____ and (b) _____ of a gas sample maintained at constant (c) _____. A mathematical expression of Boyle's Law is (d) _____. Charles' Law describes the behavior of a gas in terms of the (e) _____ and (f) _____ of a sample maintained at constant (g) _____. A mathematical expression of Charles' Law is (h) _____. The temperature in this expression must be expressed in (i) _____. Combination of (j) _____ law and (k) _____ law gives the (l) _____ gas law which provides a relationship between the (m) _____, (n) _____ and (o) _____ of a gas sample. A mathematical expression of this law is (p) _____.

Section 6.8 The Ideal Gas Law

According to Avogadro's law, (a) _____ volumes of gases at the same temperature and pressure contain (b) _____ numbers of molecules of gas. The standard conditions adopted for gas measurements are a pressure of (c) _____ _____ and a temperature of (d) _____ kelvins. A combination of Boyle's, Charles' and Avogadro's law leads to the (e) _____ _____ _____, which is written mathematically as (f) _____. The R found in the ideal gas law is called the (g) _____ gas constant.

Section 6.9 Dalton's Law

Dalton's Law is also called the law of (a) _____ _____. According to this law, the (b) _____ pressure of a mixture of gases is equal to the sum of the (c) _____ _____ of the gases in the mixture.

Section 6.10 Graham's Law

Graham's Law describes the processes of (a) _____ and (b) _____ for gases. According to this law, gases of lower molecular weight diffuse (c) _____ than gases of higher molecular weight.

Section 6.11 Changes in State

(a) _____ and (b) _____ are the processes used most often to change matter from one state to another. Such changes are called (c) _____ when heat is released or taken away, and (d) _____ when heat is added to cause the change to occur.

Section 6.12 Evaporation and Vapor Pressure

(a) _____, a process sometimes called (b) _____, is an (c) _____ process in which particles leave the surface of a liquid. The reverse process is called (d) _____. The pressure exerted by a vapor in equilibrium with liquid is called the (e) _____ _____ of the liquid.

Section 6.13 Boiling and the Boiling Point

The (a) _____ _____ of a liquid is the (b) _____ at which the liquid vapor pressure is equal to the atmospheric pressure above the liquid. A (c) _____ or (d) _____ boiling point is the temperature at which the liquid vapor pressure is equal to one (e) _____ _____.

Section 6.14 Sublimation and Melting

The process in which a solid changes directly to a (a) _____ without first becoming a liquid is called (b) _____. The temperature at which solid and liquid forms of a substance have the same vapor pressure is called the (c) _____ _____ of the substance.

Section 6.15 Energy and the States of Matter

Substances not decomposed by heating undergo a temperature change or a change in state as heat is added. The amount of heat needed to change the temperature of a specified amount of substance by 1°C is called the (a) _____ _____ of the substance. The amount of heat needed to melt one gram of a solid to a liquid at constant temperature is called the (b) _____ _____ _____ for the solid. The amount of heat needed to change one gram of liquid to a gas at constant temperature is called the (c) _____ _____ _____.

76   CHAPTER 6

SOLUTIONS TO EXERCISES ANSWERED IN THE TEXT

6.1   b) The density of liquids is obtained by dividing the mass in grams by the volume in milliliters.

$$d = \frac{(925 \text{ g})}{(500 \text{ mL})} = 1.85 \text{ g/mL}$$

d) The density of solids is obtained by dividing the mass in grams by the volume in milliliters.

$$d = \frac{(350 \text{ g})}{(200 \text{ mL})} = 1.75 \text{ g/mL}$$

6.2   a) The volume of a cube is equal to the product of the three sides, or, since all sides are equal, the cube of one side. Thus,

$$V = (\text{side})^3 = (15.0 \text{ cm})^3 = 3.38 \times 10^3 \text{ cm}^3$$

We remember that 1 cm^3 is equal to 1 mL and calculate the density as follows, after converting the 9.11 kg mass to 9.11 x 10^3 g.

$$d = \frac{(9.11 \times 10^3 \text{ g})}{(3.38 \times 10^3 \text{ mL})} = 2.70 \text{ g/mL}$$

6.3   b) The volume of the shot is obtained by subtracting the two volume readings of the graduated cylinder: V = 21.7 mL - 16.3 mL = 5.4 mL. The density is then calculated by dividing the mass in grams by the volume in milliliters.

$$d = \frac{(61.0 \text{ g})}{(5.4 \text{ mL})} = 11 \text{ g/mL}$$

6.5   The volume of the figurine is determined by subtracting the two volume readings of the graduated cylinder: V = 40.1 mL - 12.6 mL = 27.5 mL. The density is calculated by dividing the mass in grams by the volume in milliliters.

$$d = \frac{(240.8 \text{ g})}{(27.5 \text{ mL})} = 8.76 \text{ g/mL}$$

Since the density of the figurine is less than the density of silver, it can be concluded that the figurine is not made of pure silver.

The States of Matter 77

6.7 Density is calculated using the equation d = m/V. This equation can be solved for m: m = d x V. Thus, the mass of a sample of material can be determined by multiplying the volume by the density.

$$m = d \times V = 1.49 \text{ g/mL} \times 250 \text{ mL} = 373 \text{ g}$$

6.9 a) The density of both the solid and liquid are calculated by using the equation d = m/V. Since the density of the liquid is greater than the density of the solid, and a sample of solid does not change mass by melting, it must be concluded that the volume of the liquid must be less than the volume of the solid in order for the density to be greater. Thus, solid gallium contracts when it melts.
b) The density of the solid is used to calculate the mass of the solid sample.

$$m = V \times d = 5.0 \text{ mL} \times 5.90 \text{ g/mL} = 29.5 \text{ g}$$

This mass and the density of the liquid are then used to calculate the volume of the liquid as follows:

$$V = \frac{m}{d} = \frac{(29.5 \text{ g})}{(6.10 \text{ g/mL})} = 4.8 \text{ mL}$$

This volume is compared to the 5.0 mL volume of the solid and it is concluded that the volume of the liquid is 0.2 mL less than the volume of the solid before it melted.

6.11 The solution is given as the answer in Appendix B of the text.

6.13 The kinetic energy of each player is calculated using the formula

$$KE = \tfrac{1}{2}mv^2$$

Halfback: $KE = \tfrac{1}{2}mv^2 = \tfrac{1}{2}(81.8 \text{ kg})(8.0 \text{ m/s})^2 = \tfrac{1}{2}(81.8 \text{ kg})(64.0 \text{ m}^2/\text{s}^2)$

$$= 2.6 \times 10^3 \text{ kg m}^2/\text{s}^2$$

Tackle: $KE = \tfrac{1}{2}mv^2 = \tfrac{1}{2}(118.2 \text{ kg})(3.0 \text{ m/s})^2 = \tfrac{1}{2}(118.2 \text{ kg})(9.0 \text{ m}^2/\text{s}^2)$

$$= 5.3 \times 10^2 \text{ kg m}^2/\text{s}^2$$

Since the halfback has more kinetic energy, the tackle will be the one that will be pushed back.

6.15 The calculation and explanation are given as the answer in Appendix B of the text.

78 CHAPTER 6

6.16 b, d, f. The explanations are given as answers in Appendix B of the text.

6.18
b) Gases have an indefinite shape, but their densities are low. Only liquids fit this description.
d) Cohesive forces hold the particles of matter together. If cohesive forces dominate, the matter in question would have little ability to disperse (as in a gas) or flow (as in a liquid). Thus, this condition describes solids.
f) The particles of solids are less free to move about than those of liquids or gases. This characteristic describes solids.

6.20 The factors used in the calculation in each case were obtained from information given in Table 6.3.

a) The known or given quantity is 24.8 in. Hg, and the unit of the unknown quantity is atm.

$$24.8 \text{ in. Hg} \times \frac{1 \text{ atm}}{29.9 \text{ in. Hg}} = 0.829 \text{ atm}$$

c) The known or given quantity is 24.8 in Hg, and the unit of the unknown quantity is psi.

$$24.8 \text{ in. Hg} \times \frac{1 \text{ atm}}{29.9 \text{ in. Hg}} \times \frac{14.7 \text{ psi}}{1 \text{ atm}} = 12.2 \text{ psi}$$

6.21 The factors used in the calculation in each case were obtained from information given in Table 6.3.

b) The known or given quantity is 350 psi, and the unit of the unknown quantity is bar.

$$350 \text{ psi} \times \frac{1 \text{ atm}}{14.7 \text{ psi}} \times \frac{1.01 \text{ bar}}{1 \text{ atm}} = 24.0 \text{ bar}$$

d) The known or given quantity is 350 psi, and the unit of the unknown quantity is in. Hg.

$$350 \text{ psi} \times \frac{1 \text{ atm}}{14.7 \text{ psi}} \times \frac{29.9 \text{ in. Hg}}{1 \text{ atm}} = 712 \text{ in. Hg}$$

6.22 The factors used in the calculation in each case were obtained from information given in Table 6.3.

b) The known or given quantity is 17.6 cm Hg, and the unit of the unknown is mm Hg.

$$17.6 \cancel{cm} \ Hg \times \frac{10 \ mm}{1 \ \cancel{cm}} = 176 \ mm \ Hg$$

d) The known or given quantity is 17.6 cm Hg, and the unit of the unknown is psi.

$$17.6 \cancel{cm \ Hg} \times \frac{10 \ \cancel{mm}}{1 \ \cancel{cm}} \times \frac{1 \ \cancel{atm}}{760 \ \cancel{mm \ Hg}} \times \frac{14.7 \ psi}{1 \ \cancel{atm}} = 3.40 \ psi$$

6.23 b) The equation °C = K - 273 will be used.

$$°C = K - 273 = 14.1K - 273 = -258.9°C$$

d) The equation °C = K - 273 will be used.

$$°C = K - 273 = 1337.4K - 273 = 1064.4°C$$

f) The equation °C = K - 273 will be used.

$$°C = K - 273 = 505K - 273 = 232°C$$

6.25 The combined gas law will be used to solve this problem. The quantities are first identified: P_i = 690 torr, V_i = 200 mL, T_i = 26.0°C = 299 K, P_f = 760 torr, V_f = ?, T_f = 0°C = 273 K. The combined gas law is solved for the unknown, V_f, to give:

$$V_f = \frac{P_i V_i T_f}{P_f T_i} = \frac{(690 \ \cancel{torr})(200 \ mL)(273 \cancel{K})}{(760 \ \cancel{torr})(299 \cancel{K})} = 166 \ mL$$

The answer has been rounded to the correct number of significant figures.

6.27 The combined gas law will be used in the form it takes when temperature is kept constant (see Study Skills 6.1). The combined gas law at constant temperature becomes: $P_i V_i = P_f V_f$. The quantities are identified: P_i = 1 atm = 14.7 psi, V_i = ?, P_f = 65.0 psi, V_f = 1.00 L. The combined gas law is solved for the unknown, V_i, to give:

$$V_i = \frac{P_f V_f}{P_i} = \frac{(65.0 \ \cancel{psi})(1.00 \ L)}{14.7 \ \cancel{psi}} = 4.42 \ L$$

Note that the two pressures had to be expressed in the same unit so the units canceled in the calculation.

80   CHAPTER 6

6.29   The combined gas law will be used in the form it takes when the pressure is kept constant (see Study Skills 6.1). The combined gas law at constant pressure becomes:

$$\frac{V_i}{T_i} = \frac{V_f}{T_f}$$

The quantities are identified: V_i = 3.0 L, T_i = 0.0°C = 273 K, V_f = ?, T_f = 85°C = 358 K. The combined gas law is solved for the unknown, V_f, to give:

$$V_f = \frac{V_i T_f}{T_i} = \frac{(3.0\ L)(358\ K)}{(273\ K)} = 3.9\ L$$

6.31   The combined gas law will be used in the form it takes when the pressure is kept constant (see Study Skills 6.1). The combined gas law at constant pressure becomes:

$$\frac{V_i}{T_i} = \frac{V_f}{T_f}$$

The quantities are identified: V_i = ?, T_i = 100°C = 373 K, V_f = 1.0 L, T_f = 27°C = 300 K. The combined gas law is solved for the unknown, V_i, to give:

$$V_i = \frac{V_f T_i}{T_f} = \frac{(1.0\ L)(373\ K)}{(300\ K)} = 1.2\ L$$

6.33   The combined gas law will be used in the form it takes when the temperature is kept constant (see Study Skills 6.1). The combined gas law at constant temperature becomes: $P_i V_i = P_f V_f$. The quantities are identified: P_i = 1.00 atm, V_i = 2500 L, P_f = ?, V_f = 20.0 L. The combined gas law is solved for the unknown, P_f, to give:

$$P_f = \frac{P_i V_i}{V_f} = \frac{(1.00\ atm)(2500\ L)}{(20.0\ L)} = 125\ atm$$

6.35   The combined gas law will be used in its complete form. Identify the quantities: P_i = 0.98 atm, V_i = 8000 ft³, T_i = 23°C = 300 K, P_f = 400 torr = 0.526 atm, V_f = ?, T_f = 5.3°C = 278 K. The combined gas law is solved for the unknown, V_f, to give:

$$V_f = \frac{P_i V_i T_f}{P_f T_i} = \frac{(0.98\ atm)(8000\ ft^3)(278\ K)}{(0.526\ atm)(300\ K)} = 1.4 \times 10^4\ ft^3$$

Note, the two pressures had to be expressed in the same unit so the units would cancel in the calculation. The answer has been rounded to the correct number of significant figures.

6.37 The combined gas law will be used in the form it takes when the temperature is kept constant (see Study Skills 6.1). The combined gas law at constant temperature becomes: $P_iV_i = P_fV_f$. The quantities are identified: $P_i = 1.85$ atm, $V_i = 225$ mL, $P_f = 65.0$ kPa $= 0.644$ atm, V_f ?. The combined gas law is solved for the unknown, V_f, to give:

$$V_f = \frac{P_iV_i}{P_f} = \frac{(1.85 \text{ atm})(225 \text{ mL})}{(0.644 \text{ atm})} = 646 \text{ mL}$$

Note that the two pressures had to be expressed in the same unit so the units canceled in the calculation.

6.39 In order to calculate the density at the new conditions, the volume at the new conditions must first be determined. The combined gas law will be used in the form it takes when the temperature is kept constant (see Study Skills 6.1). The combined gas law at constant temperature becomes: $P_iV_i = P_fV_f$. The quantities are identified: $P_i = 760$ torr $= 1.00$ atm, $V_i = 2.00$ L, $P_f = 3.00$ atm, $V_f = ?$. The combined gas law is solved for the unknown, V_f, to give:

$$V_f = \frac{P_iV_i}{P_f} = \frac{(1.00 \text{ atm})(2.00 \text{ L})}{(3.00 \text{ atm})} = 0.667 \text{ L}$$

The density is calculated by dividing the mass of the sample by the new volume:

$$d = \frac{m}{V} = \frac{(2.50 \text{ g})}{(0.667 \text{ L})} = 3.75 \frac{g}{L}$$

6.40 In each case, the ideal gas law will be used in the form: $PV = nRT$. The value of R that will be used is 0.0821 L atm/mol K. This means that all pressures used in the equation must be expressed in atm, all volumes must be expressed in liters, and all temperatures must be expressed in kelvins.

a) The unknown quantity is P. The other quantities must be expressed in the correct units: $V = 500$ mL $= 0.500$ L, $T = 20.0°C = 293$ K, and $n = 2.00$ mol. The ideal gas law is solved for P, the unknown, to give:

$$P = \frac{nRT}{V} = \frac{(2.00 \text{ mol})(0.0821 \frac{\text{L atm}}{\text{mol K}})(293 \text{ K})}{(0.500 \text{ L})} = 96.2 \text{ atm}$$

c) The unknown quantity is T (converted to Celsius). The other quantities must be expressed in the correct units: $P = 300$ torr $= 0.395$ atm, $V = 2.50$ L, and $n = 0.100$ mol. The ideal gas law is solved for T, the unknown, to give:

82 CHAPTER 6

$$T = \frac{PV}{nR} = \frac{(0.395 \text{ atm})(2.50 \text{ L})}{(0.100 \text{ mol})(0.0821 \frac{L \cdot atm}{mol \cdot K})} = 120 \text{ K}$$

This temperature is converted to Celsius by subtracting 273.

$$T = 120K - 273 = -153°C$$

e) The unknown quantity is P. The other quantities must be expressed in the correct units: V = 2.84 L, T = 30°C = 303 K, n = 0.912 mol. The ideal gas law is solved for P, the unknown, to give:

$$P = \frac{nRT}{V} = \frac{(0.912 \text{ mol})(0.0821 \frac{L \cdot atm}{mol \cdot K})(303 K)}{(2.84 \text{ L})} = 7.99 \text{ atm}$$

6.42 The amount of gas must first be converted to moles in order to be able to use the ideal gas law.

$$n = 10.0 \text{ g } SO_2 \times \frac{1 \text{ mol } SO_2}{64.1 \text{ g } SO_2} = 0.156 \text{ mol } SO_2$$

The unknown quantity is the pressure, and the other quantities must be expressed in the correct units: V = 0.75 L, T = 27°C = 300 K, and n (calculated above) = 0.156 mol. The ideal gas law is solved for P, the unknown, to give:

$$P = \frac{nRT}{V} = \frac{(0.156 \text{ mol})(0.0821 \frac{L \cdot atm}{mol \cdot K})(300 K)}{(0.75 \text{ L})} = 5.12 \text{ atm}$$

6.44 The number of moles of methane will first be calculated using the ideal gas law, then the moles will be converted to grams. The unknown quantity is n. The other quantities must be expressed in the correct units: P = 337 psi = 22.9 atm, V = 0.50 L, T = 25°C = 298 K. The ideal gas law is solved for n, the unknown, to give:

$$n = \frac{PV}{RT} = \frac{(22.9 \text{ atm})(0.50 \text{ L})}{(0.0821 \frac{L \cdot atm}{mol \cdot K})(298 K)} = 0.47 \text{ mol}$$

The moles are converted to grams, using the molecular weight of methane, 16.04 amu:

The States of Matter   83

$$0.47 \text{ mol CH}_4 \times \frac{16.04 \text{ g CH}_4}{1 \text{ mol CH}_4} = 7.5 \text{ g CH}_4$$

6.46   The unknown quantity is the number of moles of oxygen, n. The other quantities must be expressed in the correct units: P = 1.00 atm (standard pressure), V = 60 L, and T = 273 K (standard temperature). The ideal gas law is solved for n, the unknown, to give:

$$n = \frac{PV}{RT} = \frac{(1.00 \text{ atm})(60 \text{ L})}{(0.0821 \frac{\text{L atm}}{\text{mol K}})(273 \text{ K})} = 2.7 \text{ mol}$$

6.48   The ideal gas law will be used in the form that includes the molecular weight of a gas (Equation 6.12). The unknown quantity is the molecular weight, MW. The other quantities must be expressed in the correct units: P = 1.00 atm (standard pressure), V = 3.96 L, T = 273 K (standard T), and m (the sample mass) = 8.12 g. The ideal gas law is solved for MW, the unknown, to give:

$$MW = \frac{mRT}{PV} = \frac{(8.12 \text{ g})(0.0821 \frac{\text{L atm}}{\text{mol K}})(273 \text{ K})}{(1.00 \text{ atm})(3.96 \text{ L})} = 46.0 \text{ g/mol}$$

Thus, the molecular weight of the gas is 46.0 amu.

6.50   The ideal gas law will be used in the form that includes the molecular weight of a gas (Equation 6.12). The unknown quantity is the molecular weight, MW. The other quantities must be expressed in the correct units: P = 640 torr = 0.842 atm, V = 173.6 mL = 0.1736 L, T = 20°C = 293 K, and m (the sample mass) = 0.176 g. The ideal gas law is solved for MW, the unknown, to give:

$$MW = \frac{mRT}{PV} = \frac{(0.176 \text{ g})(0.0821 \frac{\text{L atm}}{\text{mol K}})(293 \text{ K})}{(0.842 \text{ atm})(0.1736 \text{ L})} = 29.0 \frac{\text{g}}{\text{mol}}$$

Thus, the molecular weight is 29.0 amu. This most closely matches the molecular weight of CO (28 amu).

6.52   According to Dalton's law, the total pressure in the tank is equal to the sum of the partial pressures of the gases contained in the mixture in the tank. Thus, the partial pressure of CO_2 should be the total pressure minus the partial pressures of nitrogen and oxygen: $P_{carbon\ dioxide} = P_{total} - P_{nitrogen} - P_{oxygen}$ = 2200 torr - 520 torr - 650 torr = 1030 torr.

84   CHAPTER 6

6.54   According to Graham's law, the ratio of the diffusion rates is equal to the square root of the inverse ratio of the molecular weights:

$$\frac{\text{rate He}}{\text{rate HCl}} = \sqrt{\frac{\text{mass of HCl}}{\text{mass of He}}} = 3$$

The ratio of the left side of the equation is equal to 3, so the right side must be equal to the square root of 9, which is also 3. Thus, we conclude that the molecular mass of HCl must be 9 times the molecular mass of He.

6.56   The answer given in Appendix B of the text is the explanation.

6.58   Endothermic processes absorb heat, so they take place when heat is provided. Exothermic processes give up heat, so they take place when heat is removed.
b) Sublimation occurs when a solid is heated, so sublimation is endothermic.
d) Condensation occurs when a vapor is cooled, so condensation is an exothermic process.
f) Boiling occurs when a liquid is heated, so boiling is an endothermic process.

6.60   The answer in Appendix B of the text provides the explanation.

6.62   The answer in Appendix B of the text provides the explanation.

6.64   The answer in Appendix B of the text provides the explanation.

6.66   The heat required is given by the following equation:

Heat = (sample mass)(specific heat of sample)(temp. change)

Heat = (2500 g)(0.57 cal/g°C)(5°C) = 7.1 x 10^3 cal

This problem can also be solved using the specific heat given in joules per gram degree:

Heat = (2500 g)(2.4 J/g°C)(5°C) = 3.0 x 10^4 J

6.67   In each case the heat required to melt the solid (the heat that is stored) is given by the following equation:

Heat = (sample mass)(heat of fusion)

The sample mass and heat of fusion must be expressed in appropriate units so all units except those of energy cancel.

a) $\qquad$ Heat = $(1.000 \times 10^6 \text{ g})(40.7 \text{ cal/g}) = 4.07 \times 10^7$ cal

Note the sample mass was changed from 1000 kg to g in order to match the unit of the heat of fusion. This result can also be expressed in kcal or J:

$$4.07 \times 10^7 \text{ cal} \times \frac{1 \text{ kcal}}{1000 \text{ cal}} = 4.07 \times 10^4 \text{ kcal}$$

$$4.07 \times 10^7 \text{ cal} \times \frac{4.184 \text{ J}}{1 \text{ cal}} = 1.70 \times 10^8 \text{ J}$$

The conversion factor for this calculation was obtained from Table 1.3.

c) $\qquad$ Heat = $(1.000 \times 10^6 \text{ g})(57.1 \text{ cal/g}) = 5.71 \times 10^7$ cal

This result can also be changed to kcal or J as was done in part a.

$$5.71 \times 10^7 \text{ cal} \times \frac{1 \text{ kcal}}{1000 \text{ cal}} = 5.71 \times 10^4 \text{ kcal}$$

$$5.71 \times 10^7 \text{ cal} \times \frac{4.184 \text{ J}}{1 \text{ cal}} = 2.39 \times 10^8 \text{ J}$$

6.69 The heat absorbed is equal to the product of the mass of sample vaporized and the heat of vaporization.

$$\text{Heat} = (\text{sample mass})(\text{heat of vaporization})$$

$$\text{Heat} = (2000 \text{ g})(38.6 \text{ cal/g}) = 7.72 \times 10^4 \text{ cal}$$

This amount of heat can be converted to kcal and joules.

$$7.72 \times 10^4 \text{ cal} \times \frac{1 \text{ kcal}}{1000 \text{ cal}} = 7.72 \times 10^1 \text{ kcal} = 77.2 \text{ kcal}$$

$$7.72 \times 10^4 \text{ cal} \times \frac{4.184 \text{ J}}{1 \text{ cal}} = 3.23 \times 10^5 \text{ J}$$

SELF-TEST QUESTIONS

Multiple Choice

1. The density of ether is 0.736 g/mL. How much would 20.0 mL of ether weigh?
   a) 14.7 g   b) 1.47 g   c) 2.72 g   d) 27.2 g

2. Calculate the density of a swimmer who weighs 40.0 kg and occupies a volume of 45.0 liters.
   a) 0.889 g/mL
   b) 0.0088 g/mL
   c) 1.12 g/mL
   d) 0.0112 g/mL

3. As a pure liquid is heated, its temperature increases and it becomes less dense. Therefore, which of the following is true?
   a) its potential energy increases
   b) its kinetic energy increases
   c) both its potential and kinetic energy increase
   d) its potential energy increases while kinetic energy decreases

4. Which of the following is an endothermic process?
   a) freezing of water
   b) condensation of steam
   c) melting of tin
   d) solidification of liquid sulfur

5. Which of the following is a property of the gaseous state?
   a) it has a low density
   b) it has a large degree of compressibility
   c) it has a moderate degree of thermal expansion
   d) more than one response is correct

6. When the temperature is 0°C, a balloon has a volume of 5 liters. If the temperature is changed to 50°C and if we assume that the pressure inside the balloon equals atmospheric pressure at all times, which of the following will be true?
   a) the new volume will be larger
   b) the new volume will be smaller
   c) there will be no change in volume
   d) the new volume could be calculated from Boyle's law

7. A steel cylinder of CO_2 gas has a pressure of 20 atmospheres (atm) at 20°C. If the cylinder is heated to 70°C, what will the pressure be?
   a) it will be greater than 20 atm
   b) it will be less than 20 atm
   c) it will be exactly 20 atm
   d) it cannot be calculated

8. A balloon is filled with exactly 5.0 liters of helium gas in a room where the temperature is 20°C. What volume will the balloon have when it is taken outside and cooled to 10°C?
   a) 2.5 liters
   b) 4.8 liters
   c) 5.2 liters
   d) 10.0 liters

9. If the specific heat of water is 1.00 calorie per gram °C (cal/g°C) and the heat of vaporization is 540 cal/g, how much heat would it take to raise the temperature of 10.0 g of water 10°C?
   a) 10 cal
   b) 100 cal
   c) 5400 cal
   d) 54,000 cal

10. The heat of fusion of water is 80.0 cal/g. Therefore, what happens when 1.00 g of ice is melted?
    a) 80 calories of heat would be absorbed
    b) 80 calories of heat would be released
    c) no heat change would take place
    d) what would happen is not predictable

11. Calculate the amount of heat necessary to convert 5.00 g of water at 100°C into steam at 100°C. The heat of vaporization of water is 540 cal/g. The specific heat of steam is 0.480 cal/g°C.
    a) 24.0 cal
    b) 545 cal
    c) 2700 cal
    d) 2724 cal

12. What volume would 1 mole of gaseous CH_4 occupy at standard temperature and 0.82 atmospheres of pressure? Remember that R = 0.0821 L atm/mol K.
    a) 0.100 liter
    b) 1.00 liter
    c) 2.70 liters
    d) 27.3 liters

13. A sample of gas is found to occupy a volume of 6.80 liters at a temperature of 30.0°C and a pressure of 640 torr. How many moles of gas are in the sample? (R = 0.0821 L atm/mol K)
    a) 0.230 mole
    b) 4.33 moles
    c) 1769 moles
    d) 2.33 moles

14. A 1.20 L sample of gas is under a pressure of 15.0 atm. What volume would the sample occupy if the pressure were lowered to 5.20 atm and the temperature was kept constant?
    a) 0.289 L
    b) 0.416 L
    c) 3.46 L
    d) 34.6 L

True-False

15. According to Graham's law, a gas with molecules four times as heavy as a second gas will diffuse two times faster than the second gas.

16. If the volume of a gas is held constant, increasing the temperature would result in no increase in pressure.

17. The total pressure of a sample of oxygen saturated with water vapor is equal to the partial pressure of the oxygen plus the partial pressure of the water vapor.

18. The state of matter in which disruptive forces predominate is the gaseous state.

19. Boiling points of liquids decrease as atmospheric pressure decreases.

20. When a liquid is placed in a closed container, evaporation continues for a time and then stops.

21. When the combined gas law is used in a calculation, the gas pressures must be expressed in atm.

Matching

Match the states of matter given as responses to the following descriptions.

22. state has a high density
23. state has a large compressibility
24. cohesive forces dominate slightly
25. cohesive forces predominate over disruptive forces
26. sample takes shape of its container

a) solid
b) liquid
c) gas
d) two or more states

Match the terms given as responses to the following descriptions.

27. molten steel changes to a solid
28. water in an open container disappears after a time
29. ice on a car windshield disappears without melting

a) sublimation
b) evaporation
c) freezing

ANSWERS TO PROGRAMMED REVIEW

6.1　a) density　　b) compressibility　　c) gaseous
　　　d) thermal expansion

6.2　a) particles　　b) motion　　c) kinetic　　d) potential
　　　e) cohesive　　f) disruptive

6.3　a) cohesive　　b) disruptive　　c) high　　d) definite
　　　e) small　　f) very small

6.4　a) cohesive　　b) high　　c) indefinite　　d) small　　e) small

6.5　a) disruptive　　b) cohesive　　c) low　　d) indefinite

e) large   f) moderate

6.6 a) gas laws   b) pressure   c) atmosphere   d) torr

6.7 a) pressure   b) volume   c) temperature
d) PV = k or P = k/V   e) volume   f) temperature
g) pressure   h) V = k′T or V/T = k′   i) kelvins   j) Boyle's
k) Charles'   l) combined   m) pressure   n) volume
o) temperature

p) $PV/T = k''$ or $\dfrac{P_i V_i}{T_i} = \dfrac{P_f V_f}{T_f}$

6.8 a) equal   b) equal   c) one atmosphere   d) 273
e) ideal gas law   f) $PV = nRT$ g) universal

6.9 a) partial pressures   b) total   c) partial pressures

6.10 a) effusion   b) diffusion   c) faster

6.11 a) heating   b) cooling   c) exothermic   d) endothermic

6.12 a) evaporation   b) vaporization   c) endothermic
d) condensation   e) vapor pressure

6.13 a) boiling point   b) temperature   c) normal   d) standard
e) standard atmosphere

6.14 a) gas   b) sublimation   c) melting point

6.15 a) specific heat   b) heat of fusion   c) heat of vaporization

ANSWERS TO SELF-TEST QUESTIONS

1.	a	11.	d	21.	F
2.	a	12.	d	22.	d
3.	c	13.	a	23.	c
4.	c	14.	c	24.	b
5.	d	15.	F	25.	a
6.	a	16.	F	26.	d
7.	a	17.	T	27.	c
8.	b	18.	T	28.	b
9.	b	19.	T	29.	a
10.	a	20.	T		

CHAPTER 7

Solutions and Colloids

PROGRAMMED REVIEW

Section 7.1 Physical States of Solutions

Solutions are (a) _____ mixtures in which the components are present as (b) _____, (c) _____ or (d) _____. The most abundant substance in a solution is called the (e) _____, and any other substances are called (f) _____.

Section 7.2 Solubility

Substances that dissolve to a significant extent in a solvent are called (a) _____ substances. (b) _____ substances do not dissolve significantly in a solvent. The maximum amount of solute that can be dissolved in a specific amount of solvent under specific conditions is called the (c) _____ of the solute. A (d) _____ solution contains the maximum amount of dissolved solute that is stable under the prevailing conditions. An unstable solution that contains more dissolved solute than the solute solubility is called a (e) _____ solution.

Section 7.3 The Solution Process

An ion that has broken away from a solid lattice and is in solution surrounded by water molecules is called a (a) _____ ion. A saturated solution represents an (b) _____ between dissolved and undissolved solute in which undissolved solute enters solution at the same rate dissolved solute leaves solution. A good solubility generalization that applies well to nonionic compounds is (c) _____ _____ _____. One solubility rule for ionic compounds in water is that all nitrates are (d) _____. The rate of dissolving should not be confused with a solute's (e) _____. The rate of dissolving is influenced by a number of factors such as the temperature of the (f) _____.

Section 7.4 Heats of Solution

Heat is usually (a) _____ or (b) _____ when a solute dissolves in a solvent. When heat is absorbed the solution process is (c) _____, and the solution (d) _____

93

94   CHAPTER 7

as the process takes place. When heat is released, the solution process is (e) _____, and the solution temperature (f) _____ as the process takes place.

Section 7.5   Solution Concentrations

Relationships between the quantities of solute and solvent in solutions are called (a) _____. A concentration given as a (b) _____ relates the number of moles of solute in each (c) _____ of solution. In general, a concentration in (d) _____ gives the number of parts of solute contained in 100 parts of solution. The mass of solute in 100 mass units of solution is called a (e) _____/_____ percent. A weight/volume percent gives the grams of solute in (f) _____ mL of solution. A volume/volume percent is a useful concentration unit used when both the solvent and solute are either (g) _____ or (h) _____.

Section 7.6   Solution Preparation

Solutions are usually prepared by mixing together proper amounts of (a) _____ and (b) _____, or by diluting a (c) _____ solution with (d) _____ to produce a solution of lower concentration.

Section 7.7   Solution Properties

Solutes that produce water solutions that conduct (a) _____ are called (b) _____. Solutes that produce nonconductive solutions are called (c) _____. Solution properties that depend only on the concentration of solute particles are called (d) _____ properties. Three of these properties that are related are solution (e) _____ pressure, (f) _____ point and (g) _____ point. A fourth property that involves (h) _____ membranes is called (i) _____ pressure.

Section 7.8   Colloids

Colloids are (a) _____ mixtures of two or more components in which the terms (b) _____ medium and (c) _____ phase are used in a manner analogous to the terms solvent and solute for solutions. The (d) _____ effect is a property of colloids in which the path of a light beam passing through the colloid is visible.

Section 7.9   Types of Colloids

The word "colloidal" means (a) _____, and some colloids fit that description. However, colloids are usually differentiated according to the physical (b) _____ of the dispersing medium and dispersed phase. In a colloid called a foam, the dispersing medium is a (c) _____ and the dispersed phase is a (d) _____.

Solutions and Colloids 95

Section 7.10 Colloid Formation and Destruction

Much of the interest in colloids is related to their (a) _____ or (b) _____. In a Cottrel precipitator, colloidal (c) _____ are removed from (d) _____ smoke stack waste. Some colloids are stabilized by substances called (e) _____ or (f) _____ agents.

Section 7.11 Dialysis

(a) _____ membranes are semipermeable membranes that hold back (b) _____ particles and large (c) _____, but allow solvent, hydrated ions and small molecules to pass through. The passage of the ions and small molecules is called (d) _____, a process used to cleanse the blood of people suffering (e) _____ malfunction.

SOLUTIONS TO EXERCISES ANSWERED IN THE TEXT

7.1 b) No water is mentioned, but it is the likely solvent. The solutes would be alcohol and opium.
   d) No water is mentioned, but it is the likely solvent. The solute is acetic acid.
   f) No water is mentioned, but it is a likely component of the mixture. However, the water could have a maximum percentage of only 30 %, so it would be listed as a solute rather than a solvent. The solvent is isopropyl alcohol.

7.2 The answers in Appendix B of the text include the explanations.

7.3 b) The cloudiness of the mixture indicates that a good part of the solid did not dissolve. Thus, the solid is classified as insoluble.
   d) The fact that the resulting mixture is clear indicates that the butter has dissolved in the chloroform. The yellow color is not an indication of insolubility, it simply indicates that some yellow-colored components of the butter dissolved in the chloroform.
   f) The presence of solid iodine on the bottom of the container indicates that most of the iodine did not dissolve. The slight brown color of the liquid indicates that a slight amount of the solid iodine dissolved, but the large amount of solid that did not dissolve would lead to a classification of insoluble for the solid iodine.

7.5 a) The fact that the added solid dissolved indicates that the solution still has the capacity to dissolve more solute. This means the solution is unsaturated.

96 CHAPTER 7

c) The original solution in part b would be classified as supersaturated because some dissolved solute was released. The solution that remains after a supersaturated solution has given up its excess solute is saturated.

7.6   According to the information in Table 7.2, 100 g of water at 0°C has the capacity to dissolve 70.6 g of ammonium sulfate. Since only 35.8 g of ammonium sulfate was added to the 100 g of water, it should all dissolve and produce a solution that could still dissolve more. Such a solution is unsaturated.

7.8   The answers given in Appendix B of the text give the classifications based on the categories of Table 7.3 and the solubility information for the solutes.

7.9   The answer given in Appendix B of the text provides the explanation and drawing.

7.11 The answer given in Appendix B of the text provides the explanation.

7.12 In each case, the solute will dissolve in the solvent that it most resembles in terms of polarity.
  b) Neon molecules are single atoms and cannot be polar. Neon would therefore tend to dissolve in a nonpolar solvent such as benzene.
  d) The boron trifluoride molecule as shown is symmetrical and so is not polar. It would tend to dissolve in a nonpolar solvent such as benzene.
  f) Hydrogen chloride is a polar molecule and would tend to dissolve in a polar solvent such as water.

7.14 The solid could be crushed to increase the amount of surface area of solute that is available for attack by the solvent. The solution could be heated to increase the kinetic energy of the solvent molecules and make them more effective in separating dissolved solid from the undissolved solid. The solution could be stirred to remove saturated or more concentrated solution from the vicinity of the dissolving solid, and expose the solid to fresh solvent or less concentrated solution.

7.15 The molarity of a solution is obtained by dividing the number of moles of solute dissolved in a specific volume of the solution by the specific volume. The solution volume must always be expressed in liters. In this case, 0.112 moles of solute is dissolved in a volume of 250 mL or 0.250 L of solution. The molarity is calculated as follows:

$$\text{molarity} = \frac{\text{moles of solute}}{\text{liters of solution}} = \frac{(0.112 \text{ mol})}{(0.250 \text{ L})} = 0.448 \text{ M}$$

7.16 In each case, the molarity is calculated by dividing the number of moles of solute dissolved in a specific volume of solution by the volume of solution, with the solution volume expressed in liters. If the mass of solute dissolved in a specific volume is given, the mass must be converted to moles.

b) The 20.2 g of solid $CuCl_2$ must be converted to moles:

$$20.2 \text{ g } CuCl_2 \times \frac{1 \text{ mol } CuCl_2}{134.5 \text{ g } CuCl_2} = 0.150 \text{ mol } CuCl_2$$

The calculated number of moles of solute is now divided by the liters of solution in which it is dissolved to obtain the molarity.

$$\text{molarity} = \frac{\text{mol of solute}}{\text{liters of solution}} = \frac{(0.150 \text{ mol})}{(1.00 \text{ L})} = 0.150 \text{ M}$$

d) The 0.140 g of solid Na_2SO_4 must be converted to moles:

$$0.140 \text{ g } Na_2SO_4 \times \frac{1 \text{ mol } Na_2SO_4}{142.0 \text{ g } Na_2SO_4} = 9.86 \times 10^{-4} \text{ mol } Na_2SO_4$$

The calculated number of moles of solute is now divided by the liters of solution in which it is dissolved.

$$\text{molarity} = \frac{\text{mol of solute}}{\text{liters of solution}} = \frac{(9.86 \times 10^{-4} \text{ mol})}{(0.010 \text{ L})} = 0.0986 \text{ M}$$

f) In this exercise, the volume of solution that contains the 43.5 g of solute is not known. However, it is known that 2.18 g of solute was contained in a 25.0 mL sample of the solution. The 2.18 g of K_2SO_4 is first converted to moles:

$$2.18 \text{ g } K_2SO_4 \times \frac{1 \text{ mol } K_2SO_4}{174.3 \text{ g } K_2SO_4} = 0.0125 \text{ mol } K_2SO_4$$

The calculated number of moles of solute is now divided by the liters of solution in which it is dissolved.

$$\text{molarity} = \frac{\text{mol of solute}}{\text{liters of solution}} = \frac{(0.0125 \text{ mol})}{(0.0250 \text{ L})} = 0.500 \text{ M}$$

7.17 In each case, the pathway given by Figure 7.8 will be followed.
 b) This problem has the pattern mol A → liters soln. A, and the pathway is mol HNO_3 → liters HNO_3 solution. The necessary factor comes from the solution molarity.

98 CHAPTER 7

$$0.0410 \text{ mol HNO}_3 \times \frac{1 \text{ L HNO}_3 \text{ soln}}{0.315 \text{ mol HNO}_3} = 0.130 \text{ L HNO}_3 \text{ solution}$$

This volume can also be expressed as 130 mL HNO$_3$ solution

d) This problem has the pattern liters soln. A → mol A → grams A, and the pathway is liters AgNO$_3$ solution → mol AgNO$_3$ → grams AgNO$_3$. The necessary factors come from the solution molarity and the formula weight of AgNO$_3$.

$$0.200 \text{ L AgNO}_3 \text{ soln.} \times \frac{0.200 \text{ mol AgNO}_3}{1 \text{ L AgNO}_3 \text{ soln.}} \times \frac{169.9 \text{ g AgNO}_3}{1 \text{ mol AgNO}_3}$$

$$= 6.80 \text{ g AgNO}_3$$

f) This problem has the pattern liters soln. A → mol A, and the pathway is liters HCl solution → mol HCl. The necessary factor comes from the solution molarity.

$$0.250 \text{ L HCl soln.} \times \frac{6.0 \text{ mol HCl}}{1 \text{ L HCl soln.}} = 1.5 \text{ mol HCl}$$

7.18 In each case, the pathway given by Figure 7.8 will be followed.
 a) This problem has the pattern grams A → liters soln. B, and the pathway is grams NaOH → mol NaOH → mol HCl → liters HCl solution. The necessary factors come from the formula weight of NaOH, the equation coefficients and the HCl solution molarity.

$$25.0 \text{ g NaOH} \times \frac{1 \text{ mol NaOH}}{40.0 \text{ g NaOH}} \times \frac{1 \text{ mol HCl}}{1 \text{ mol NaOH}} \times \frac{1 \text{ L HCl soln}}{6 \text{ mol HCl}}$$

$$= 0.104 \text{ L HCl solution}$$

This can also be expressed as 104 mL HCl solution.

c) This problem has the pattern grams A → liters soln. B, and the pathway is grams NaHCO$_3$ → mol NaHCO$_3$ → mol HCl → liters HCl solution. The necessary factors come from the formula weight of NaHCO$_3$, the equation coefficients and the HCl solution molarity.

Solutions and Colloids 99

$$10.5 \text{ g NaHCO}_3 \times \frac{1 \text{ mol NaHCO}_3}{84.9 \text{ g NaHCO}_3} \times \frac{1 \text{ mol HCl}}{1 \text{ mol NaHCO}_3} \times \frac{1 \text{ L HCl soln.}}{0.250 \text{ mol HCl}}$$

$$= 0.500 \ L \ HCl \ solution$$

This can also be expressed as 500 mL HCl solution.

7.19 In each case, the pathway given by Figure 7.8 will be followed.
  a) This problem has the pattern liters soln. A → liters soln. B, and the pathway is liters NaCl solution → mol NaCl → mol AgNO$_3$ → liters AgNO$_3$ solution. The necessary factors come from the solution molarities and the equation coefficients.

$$0.025 \text{ L NaCl soln.} \times \frac{0.200 \text{ mol NaCl}}{1 \text{ L NaCl soln.}} \times \frac{1 \text{ mol AgNO}_3}{1 \text{ mol NaCl}} \times \frac{1 \text{ L AgNO}_3 \text{ soln.}}{0.250 \text{ mol AgNO}_3}$$

$$= 0.0200 \ L \ AgNO_3 \ soln.$$

This volume can also be expressed as 20.0 mL of AgNO$_3$ solution.

  c) This problem has the pattern liters soln. A → liters soln. B, and the pathway is liters H$_2$SO$_4$ solution → mol H$_2$SO$_4$ → mol NH$_3$ → liters NH$_3$ solution. The necessary factors come from the solution molarities and the equation coefficients.

$$0.0200 \text{ L H}_2\text{SO}_4 \text{ soln.} \times \frac{0.145 \text{ mol H}_2\text{SO}_4}{1 \text{ L H}_2\text{SO}_4 \text{ soln.}} \times \frac{2 \text{ mol NH}_3}{1 \text{ mol H}_2\text{SO}_4} \times \frac{1 \text{ L NH}_3 \text{ soln.}}{0.200 \text{ mol NH}_3}$$

$$= 0.0290 \ L \ NH_3 \ soln.$$

This volume can also be expressed as 29.0 mL of NH$_3$ solution.

  e) This problem has the pattern liters soln. A → liters soln. B, and the pathway is liters H$_2$SO$_4$ solution → mol H$_2$SO$_4$ → mol NaOH → liters NaOH solution. The necessary factors come from the solution molarities and the equation coefficients.

100 CHAPTER 7

$$0.0250 \text{ L } H_2SO_4 \text{ soln.} \times \frac{0.125 \text{ mol } H_2SO_4}{1 \text{ L } H_2SO_4 \text{ soln.}} \times \frac{2 \text{ mol } NaOH}{1 \text{ mol } H_2SO_4} \times \frac{1 \text{ L } NaOH \text{ soln.}}{0.108 \text{ mol } NaOH}$$

$$= 0.0579 \, L \, NaOH \, solution$$

This volume can also be expressed as 57.9 mL of NaOH.

7.21 This problem has the pattern liters soln. A → grams B, and the pathway is liters stomach acid (HCl) → mol stomach acid (HCl) → mol Mg(OH)$_2$ → grams Mg(OH)$_2$. The necessary factors come from the solution molarity (see exercise 7.20) and the equation coefficients.

$$0.250 \text{ L HCl soln.} \times \frac{0.10 \text{ mol HCl}}{1 \text{ L HCl soln.}} \times \frac{1 \text{ mol } Mg(OH)_2}{2 \text{ mol HCl}} \times \frac{58.33 \text{ g } Mg(OH)_2}{1 \text{ mol } Mg(OH)_2}$$

$$= 0.73 \, g \, Mg(OH)_2$$

If the results of working exercise 7.20 are available, it will be seen that Mg(OH)$_2$ is more efficient at reacting with stomach acid.

7.22 The %(w/w) concentration is calculated using the following equation:

$$\%(w/w) = \frac{solute \; mass}{solution \; mass} \times 100$$

b) In this problem it must be remembered to include the mass of solute in the total solution mass. Thus, the mass of solute is 4.0 grams, and the mass of the solution is the 100 g from the 100 mL of water used plus the 4.0 grams of solute, or 104 grams.

$$\%(w/w) = \frac{4.0 \text{ g solute}}{104 \text{ g soln.}} \times 100 = 3.8\%$$

7.23 In each case the %(w/w) will be calculated using the equation given in exercise 7.22 above. It is important to remember that the solution mass will include the mass of water used plus the mass of solute used.
b) The mass of 0.10 mol of glucose must be determined:

$$0.10 \text{ mol } C_6H_{12}O_6 \times \frac{180.2 \text{ g } C_6H_{12}O_6}{1 \text{ mol } C_6H_{12}O_6} = 18 \text{ g } C_6H_{12}O_6$$

The solution mass will thus be 118 g.

$$\%(w/w) = \frac{18 \text{ g solute}}{118 \text{ g soln.}} \times 100 = 15\%$$

d) The mass of 5.00 mol of water must be determined:

$$5.00 \text{ mol } H_2O \times \frac{18.0 \text{ g } H_2O}{1 \text{ mol } H_2O} = 90.0 \text{ g } H_2O$$

The solution mass will thus be 90.0 g + 5.2 g = 95.2 g

$$\%(w/w) = \frac{5.2 \text{ g solute}}{95.2 \text{ g soln.}} \times 100 = 5.5\%$$

f) The mass of each liquid is first determined by using the density formula, d = m/V. According to this formula, the mass of a liquid is equal to d x V.

Mass of ethyl alcohol = 10.0 mL x 0.789 g/mL = 7.89 g
Mass of ethylene glycol = 10.0 mL x 1.11 g/mL = 11.1 g
The total mass of the solution is thus 18.99 g. The %(w/w) will be calculated for each substance in the solution.

Ethyl alcohol: $\%(w/w) = \dfrac{7.89 \text{ g solute}}{18.99 \text{ g soln.}} \times 100 = 41.5\%$

Ethylene glycol: $\%(w/w) = \dfrac{11.1 \text{ g solute}}{18.99 \text{ g soln.}} \times 100 = 58.5\%$

7.24 b) The solution mass is obtained by using the density formula, d = m/V, or m = d x V. m = 1.10 g/ml x 10.0 mL = 11.0 g. The mass of the residue remaining after evaporation is the mass of the solute, 0.92 g.

$$\%(w/w) = \frac{0.92 \text{ g solute}}{11.0 \text{ g soln.}} \times 100 = 8.4\%$$

102 CHAPTER 7

7.25 In each case, the %(v/v) will be calculated using the following equation:

$$\%(v/v) = \frac{solute\ volume}{solution\ volume} \times 100$$

b) $\%(v/v) = \dfrac{30\ mL\ solute}{500\ mL\ soln.} \times 100 = 6.0\%$

d) $\%(v/v) = \dfrac{215\ mL\ alcohol}{500\ mL\ soln.} \times 100 = 43.0\%\ alcohol$

7.27 The volume of acetone is determined using the density formula, d = m/V, or V = m/d. It must also be noted that the mass of acetone given as 1.8 mg must be expressed in grams so the units cancel properly in the calculation.

$$V = \frac{(1.8 \times 10^{-3}\ g)}{(0.79\ g/mL)} = 2.3 \times 10^{-3}\ mL$$

$$\%(v/v) = \frac{2.3 \times 10^{-3}\ mL\ solute}{100\ mL\ soln.} \times 100 = 2.3 \times 10^{-3}\ \%$$

7.28 In each case, the %(w/v) will be calculated using the following equation:

$$\%(w/v) = \frac{g\ of\ solute}{mL\ of\ solution} \times 100,$$

where care must be taken to express the quantities in the correct units.

b) $\%(w/v) = \dfrac{5.0\ g\ solute}{250\ mL\ soln.} \times 100 = 2.0\%$

d) The solution will have a total mass of 28.0 g + 200 g = 228 g, where the 200 g comes from assuming that the density of water is 1.0 g/mL. The density of the resulting solution is 1.10 g/mL. This density and the solution mass will be used in the density equation to calculate the volume of the solution.

d = m/V, or V = m/d   Thus, the solution volume, V, is calculated as follows:

$$V = \frac{m}{d} = \frac{(228 \text{ g})}{(1.10 \text{ g/mL})} = 207 \text{ mL}$$

$$\%(w/v) = \frac{28.0 \text{ g solute}}{207 \text{ mL soln.}} \times 100 = 13.5\%$$

f) $$\%(w/v) = \frac{1.02 \text{ g solute}}{15.0 \text{ mL soln.}} \times 100 = 6.80\%$$

7.29 According to Figure 7.3, a saturated solution of KCl at 25 °C contains about 32 g of KCl per 100 g H_2O,

$$\%(w/w) = \frac{\text{solute mass}}{\text{solution mass}} \times 100 = \frac{32 \text{ g solute}}{132 \text{ g soln.}} \times 100 = 24\%$$

This answer is approximate and depends upon how carefully the information is obtained from Figure 7.3.

7.31 b) The equation defining molarity,

$$M = \frac{\text{mol of solute}}{\text{L of solution}}$$

can be used to calculate the number of moles of $AgNO_3$ contained in 100 mL (0.100 L) of solution.

mol solute = M x L soln. = 0.20 M x 0.100 L = 0.020 mol $AgNO_3$
The mass of $AgNO_3$ needed can be obtained by converting the moles of $AgNO_3$ into grams:

$$0.020 \text{ mol AgNO}_3 \times \frac{169.9 \text{ g AgNO}_3}{1 \text{ mol AgNO}_3} = 3.4 \text{ g AgNO}_3$$

This amount of solute should be put into enough distilled water to produce 100 mL of solution.

d) The equation defining %(w/v),

$$\%(w/v) = \frac{grams\ of\ solute}{mL\ of\ solution} \times 100$$

can be used to calculate the number of grams of NaCl that will be needed to make 500 mL of solution.

$$g\ solute = \frac{\%(w/v) \times mL\ of\ soln.}{100} = \frac{0.90\% \times 500\ mL}{100} = 4.5\ g\ solute$$

The 4.5 g of NaCl should be added to enough distilled water to give 500 mL of solution.

f) The equation defining %(w/v),

$$\%(w/v) = \frac{grams\ of\ solute}{mL\ of\ solution} \times 100$$

can be used to calculate the number of grams of glycerol that will be needed to make 100 mL of solution.

$$g\ solute = \frac{\%(w/v) \times mL\ of\ soln.}{100} = \frac{10\% \times 100\ mL}{100} = 10\ g\ solute$$

The 10 grams of glycerol can be weighed out and added to enough water to give 100 mL, or the volume of glycerol with a mass of 10 g can be determined from the density of glycerol, and the volume can be measured. $d = m/V$, so $V = m/d$,

$$V = \frac{(10\ g)}{(1.26\ g/mL)} = 7.9\ mL$$

Thus, 7.9 mL of glycerol contains 10 g, and could be used.

h) The equation defining %(w/w),

$$\%(w/w) = \frac{solute\ mass}{solution\ mass} \times 100$$

can be used to calculate the number of grams of KCl needed to make 100 g of solution.

$$g\ solute = \frac{\%(w/w) \times soln.\ mass}{100} = \frac{4.0\% \times 100\ g}{100} = 4.0\ g\ solute$$

The 4.0 grams of KCl should be added to enough water to give 100 g of solution. Since 1 mL of water has a mass of 1 g, 96 mL of water would be required.

7.33 b) The equation defining molarity,

$$M = \frac{mol\ of\ solute}{L\ of\ solution}$$

can be used to calculate the number of moles of KBr in the solution.

moles KBr = M × L soln. = 0.72 M × 0.120 L soln. = 0.086 mol KBr.

The moles of KBr can be converted to grams by using the formula weight of KBr.

$$0.086\ mol\ KBr \times \frac{119.0\ g\ KBr}{1\ mol\ KBr} = 10\ g\ KBr$$

d) The equation defining %(v/v),

$$\%(v/v) = \frac{solute\ volume}{solution\ volume} \times 100$$

can be used to calculate the volume of alcohol in the solution.

$$solute\ volume = \frac{\%(v/v) \times soln.\ volume}{100} = \frac{20.0\% \times 250\ mL}{100} = 50\ mL$$

f) The equation defining molarity,

$$M = \frac{mol\ solute}{L\ solution}$$

can be used to calculate the number of moles of NH_3 in the solution.

moles NH_3 = M × L soln. = 6.0 M × 0.250 L soln. = 1.5 mol NH_3

The moles of NH_3 can be converted to grams by using the molecular weight of NH_3.

$$1.5 \text{ mol NH}_3 \times \frac{17.0 \text{ g NH}_3}{1 \text{ mol NH}_3} = 26 \text{ g NH}_3$$

7.34 In each case where the concentrations of the dilute and concentrated solutions are in the same units, Equation 7.6 will be used, $(C_c)(V_c) = (C_d)(V_d)$, where the c and d subscripts refer to the concentrated and dilute solutions respectively.

b) $(6.0 \text{ M})(V_c) = (2.0 \text{ M})(50 \text{ mL})$; $V_c = (2.0 \text{ M})(50 \text{ mL})/(6.0 \text{ M})$, $V_c = 17$ mL. Thus, the volume of 6 M H_2SO_4 needed is 17 mL. Put that volume into a 50 mL flask, and add distilled water up to the mark.

d) $(3.0 \text{ M})(V_c) = (0.50 \text{ M})(250 \text{ mL})$; $V_c = (0.50 \text{ M})(250 \text{ mL})/(3.0 \text{ M})$, $V_c = 42$ mL. Thus, the volume of 3.0 M $CaCl_2$ needed is 42 mL. Put that volume into a 250 mL flask and add distilled water up to the mark.

f) $(18.0 \text{ M})(V_c) = (6.0 \text{ M})(5.0 \text{ L})$; $V_c = (6.0 \text{ M})(5.0 \text{ L})/(18.0 \text{ M})$, $V_c = 1.7$ L. Thus, 1.7 L of 18.0 M H_2SO_4 is needed. Put the 1.7 L into about 3 L of distilled water. This must be done carefully. See the answer in Appendix B of the text for details.

h) The equation defining molarity will be used to calculate the number of moles of $AgNO_3$ needed to make the 500 mL of 0.100 solution.

$$M = \frac{\text{mol solute}}{\text{L solution}}, \text{ therefore mol AgNO}_3 = M \times L \text{ soln.}$$

mol $AgNO_3$ = 0.100 M x 0.500 L soln. = 0.0500 mol $AgNO_3$
The moles of $AgNO_3$ is then converted into grams.

$$0.0500 \text{ mol AgNO}_3 \times \frac{169.9 \text{ g AgNO}_3}{1 \text{ mol AgNO}_3} = 8.50 \text{ g AgNO}_3$$

This mass of $AgNO_3$ must be obtained from the 5.00 %(w/v) solution. The equation that defines the %(w/v),

$$\%(w/v) = \frac{\text{g of solute}}{\text{mL of soln.}} \times 100$$

will be used to calculate the volume of solution needed to provide the $AgNO_3$.

$$\text{mL soln.} = \frac{\text{g of solute}}{\%(w/v)} \times 100 = \frac{8.50 \text{ g AgNO}_3}{5.00\%} \times 100 = 170 \text{ mL}$$

Thus, 170 mL of the 5.00 %(w/v) solution will have to be diluted to a final volume of 500 mL.

Solutions and Colloids   107

7.36   The answer in Appendix B of the text provides the explanation.

7.37   In each case, the equations $\Delta t_b = nK_bM$ and $\Delta t_f = nK_fM$ will be used to calculate the change in boiling point and freezing point respectively for solutions compared to pure solvents. The boiling and freezing point constants and normal boiling and freezing points for solvents were obtained from Table 7.6.

   b) $\Delta t_b = (0.52 \,°C/M)(1.00 \,M) = 0.52 \,°C$. The change is added to the normal boiling point, so the solution would boil at $100.0 \,°C + 0.52°C$, or $100.52°C$.

   $\Delta t_f = (1.86 \,°C/M)(1.00 \,M) = 1.86 \,°C$. The change is subtracted from the normal freezing point, so the solution would freeze at $0.0°C - 1.86 \,°C$, or $-1.86 \,°C$.

   d) In this problem, $n = 3$ because the $Mg(NO_3)_2$ is a strong electrolyte that will dissociate into one Mg^{2+} ion and two NO_3^- ions.

   $\Delta t_b = (3)(0.52 \,°C/M)(1.00 \,M) = 1.56 \,°C$. The change is added to the normal boiling point, so the solution would boil at $101.56 \,°C$.

   $\Delta t_f = (3)(1.86 \,°C/M)(1.00 \,M) = 5.58 \,°C$. The change is subtracted from the normal freezing point, so the solution would freeze at $-5.58 \,°C$.

   f) In this problem, $n = 5$ because $Al_2(SO_4)_3$ is a strong electrolyte that will dissociate into two Al^{3+} and three SO_4^{2-} ions.

   $\Delta t_b = (5)(0.52 \,°C/M)(1.00 \,M) = 2.60 \,°C$. The change is added to the normal boiling point, so the solution would boil at $102.60 \,°C$.

   $\Delta t_f = (5)(1.86 \,°C/M)(1.00 \,M) = 9.30 \,°C$. The change is subtracted from the normal freezing point, so the solution would freeze at $-9.30 \,°C$.

7.38   In each case, the equations for calculating the change in boiling and freezing point used in exercise 7.37 above will be used. The necessary constants and normal boiling points are from Table 7.6.

   b) $\Delta t_b = (3)(0.52 \,°C/M)(0.250 \,M) = 0.39 \,°C$. The change is added to the normal boiling point, so the solution would boil at $100.39°C$.

   $\Delta t_f = (3)(1.86 \,°C/M)(0.250 \,M) = 1.40 \,°C$. The change is subtracted from the normal freezing point, so the solution would freeze at $-1.40 \,°C$.

   d) The moles of H_2SO_4 in the solution must be calculated as well as the molarity of the solution.

$$50.0 \text{ g H}_2\text{SO}_4 \times \frac{1 \text{ mol H}_2\text{SO}_4}{98.1 \text{ g H}_2\text{SO}_4} = 0.510 \text{ mol H}_2\text{SO}_4$$

$$\text{Solution molarity, } M = \frac{0.510 \text{ mol H}_2\text{SO}_4}{0.250 \text{ L soln.}} = 2.04 \text{ M}$$

$\Delta t_b = (3)(0.52 \text{ °C/M})(2.04 \text{ M}) = 3.18 \text{ °C}$. The change is added to the normal boiling point, so the solution would boil at 103.18 °C.

$\Delta t_f = (3)(1.86 \text{ °C/M})(2.04 \text{ M}) = 11.4 \text{ °C}$. The change is subtracted from the normal freezing point, so the solution would freeze at -11.4 °C.

f) The moles of octanoic acid in the solution and the solution molarity must be calculated. Also note that the solvent is benzene rather than water.

$$75.0 \text{ g C}_8\text{H}_{16}\text{O}_2 \times \frac{1 \text{ mol C}_8\text{H}_{16}\text{O}_2}{144.2 \text{ g C}_8\text{H}_{16}\text{O}_2} = 0.520 \text{ mol C}_8\text{H}_{16}\text{O}_2$$

$$\text{Solution molarity, } M = \frac{0.520 \text{ mol C}_8\text{H}_{16}\text{O}_2}{0.250 \text{ L}} = 2.08 \text{ M}$$

$\Delta t_b = (2.53 \text{ °C/M})(2.08 \text{ M}) = 5.26 \text{ °C}$. The change is added to the normal boiling point, so the solution would boil at 80.1 + 5.3 or 85.4 °C.

$\Delta t_f = (4.90 \text{ °C/M})(2.08 \text{ M}) = 10.2 \text{ °C}$. The change is subtracted from the normal freezing point, so the solution would freeze at 5.5-10.2 or -4.7 °C.

7.39 In each case it is recognized that the osmolarity of the solution is equal to the product of n and M in the equation for osmotic pressure, $\pi = n\text{MRT}$.
b) KCl is a strong electrolyte and will dissociate into K^+ and Cl^- ions. Thus, $n = 2$. Osmolarity = 2 x 0.25 mol/L = 0.50 mol/L.
d) KOH is a strong electrolyte and will dissociate into K^+ and OH^- ions. Thus, $n = 2$. The number of moles of KOH and the molarity of the solution must be calculated:

$$45.0 \text{ g KOH} \times \frac{1 \text{ mol KOH}}{56.1 \text{ g KOH}} = 0.802 \text{ mol KOH}$$

Solutions and Colloids   109

$$\text{Solution molarity} = M = \frac{\text{mol of solute}}{\text{L of solution}} = \frac{0.802 \text{ mol KOH}}{1 \text{ L}}$$

$$= 0.802 \; M$$

Osmolarity = 2 x 0.802 mol/L = 1.60 mol/L

f) Ethylene glycol is a nonelectrolyte, so $n = 1$. The number of moles of ethylene glycol and the molarity of the solution must be calculated.

$$50.0 \text{ mL } C_2H_6O_2 \times \frac{1.11 \text{ g } C_2H_6O_2}{1 \text{ mL } C_2H_6O_2} \times \frac{1 \text{ mol } C_2H_6O_2}{62.1 \text{ g } C_2H_6O_2} = 0.894 \text{ mol } C_2H_6O_2$$

$$\text{Solution molarity} = M = \frac{0.894 \text{ mol } C_2H_6O_2}{0.250 \text{ L soln.}} = 3.58 \text{ mol/L}$$

Osmolarity = 1 x 3.58 mol/L = 3.58 mol/L

7.40 In each case, the R value will be 62.4 L torr/K mol, which will give the answer in torr. This answer will also represent the value in mm Hg. The answer will be converted to atm. by dividing the answer in torr by 760 torr/atm.

b) $Ca(NO_3)_2$ is a strong electrolyte that will dissociate into one Ca^{2+} and two NO_3^- ions. Thus, $n = 3$.

$$\pi = nMRT = 3 \times 0.100 \text{ mol/L} \times 62.4 \text{ L torr/K mol} \times 298 \text{ K}$$
$$= 5.58 \times 10^3 \text{ torr} = 5.58 \times 10^3 \text{ mm Hg} = 7.34 \text{ atm.}$$

d) Urea is a nonelectrolyte so $n = 1$. The number of moles of urea and the molarity of the solution must be determined.

$$35.0 \text{ g } CH_4N_2O \times \frac{1 \text{ mol } CH_4N_2O}{60.1 \text{ g } CH_4N_2O} = 0.583 \text{ mol } CH_4N_2O$$

$$\text{Solution molarity} = M = \frac{0.583 \text{ mol } CH_4N_2O}{0.100 \text{ L soln.}} = 5.83 \text{ mol/L}$$

$$\pi = nMRT = 1 \times 5.83 \text{ mol/L} \times 62.4 \text{ L torr/K mol} \times 298 \text{ K}$$
$$= 1.08 \times 10^4 \text{ torr} = 1.08 \times 10^4 \text{ mm Hg} = 143 \text{ atm}$$

f) According to Equation 7.11, the product of $nM = \Delta t_f/K_f$. If the solution freezing point is -0.47 °C, then Δt_f must be 0.47 °C. The value of K_f for water is 1.86 °C/M (Table 7.6). Thus, the value of
$nM = 0.47\ °C/1.86\ °C/M = 0.25$ mol/L.

$\pi = nMRT = 0.25$ mol/L $\times$ 62.4 L torr/K mol $\times$ 298 K $= 4.7 \times 10^3$ torr
$= 4.7 \times 10^3$ mm Hg $= 6.2$ atm

h) Both NaCl and KCl are strong electrolytes that produce the Na^+, Cl^-, K^+ and Cl^- ions respectively. Thus, $n = 2$ for each electrolyte, so the electrolytes can be treated as a single solute. We must determine the number of moles of each one, then the effective number of moles will be the sum of the two.

5.3 g NaCl $\times \dfrac{1\ mol\ NaCl}{58.49\ g\ NaCl} = 0.091\ mol\ NaCl$

8.2 g KCl $\times \dfrac{1\ mol\ KCl}{74.55\ g\ KCl} = 0.11\ mol\ KCl$

Moles $= 0.091 + 0.11 = 0.20$ mol/L

$M = \dfrac{0.20\ mol}{0.750\ L} = 0.27\ mol/L$

$\pi = nMRT = 2 \times 0.27$ mol/L $\times$ 62.4 L torr/K mol $\times$ 298 K $= 10 \times 10^3$ torr
$= 10 \times 10^3$ mm Hg $= 13$ atm

7.42   The answer in Appendix B of the text gives the explanation.

SELF-TEST QUESTIONS

Multiple Choice

1. A solution is prepared by dissolving a small amount of sugar in a large amount of water. In this case sugar would be the
   a) filtrate
   b) solute
   c) precipitate
   d) solvent

2. The freezing point of a solution
   a) is lower than that of pure solvent
   b) cannot be measured
   c) is higher than that of pure solvent
   d) is the same as that of pure solvent

3. Which of the following would show the Tyndall effect?
   a) a solution of salt in water
   b) a solution of sugar in water
   c) a solution of CO_2 gas in water
   d) a colloidal suspension

4. A crystal of solid magnesium sulfate is placed into a solution of magnesium sulfate in water. It is observed that the crystal dissolves slightly. The original solution was
   a) saturated
   b) unsaturated
   c) supersaturated
   d) cannot determine from the data given

5. The weight/weight percent of sugar in a solution containing 25 grams of sugar and 75 grams of water would be:
   a) 25%   b) 33%   c) 75%   d) 80%

6. How many moles of $Mg(NO_3)_2$ is contained in 500 mL of 0.400 M solution?
   a) 0.400   b) 0.200   c) 0.800   d) 2.00

7. How many grams of $Mg(NO_3)_2$ must be dissolved in water to give 500 mL of 0.400 M solution?
   a) 29.7   b) 17.3   c) 59.3   d) 172.6

8. Which of the following aqueous solutions would be expected to have the highest boiling point at 1 atm pressure?
   a) 1 M NaCl
   b) 1 M $C_{12}H_{22}O_{11}$ (sucrose)
   c) pure water
   d) 1 M AlF_3

9. Calculate the freezing point of a water solution that contains 1.60 grams of methyl alcohol (a nonelectrolyte), CH_3OH, in each 100.0 mL. The K_f for water is 1.86°C/M.
   a) 0.93°C   b) -0.93°C   c) -1.86°C   d) -3.72°

10. How many moles of solute would be needed to prepare 250 mL of 0.150 M solution?
    a) 37.5   b) 0.150   c) 0.0375   d) 0.600

11. What volume of 0.200 M silver nitrate solution ($AgNO_3$) would have to be diluted to form 500 mL of 0.050 M solution?
    a) 200 mL   b) 2.0×10^{-5} mL   c) 80.0 mL   d) 125 mL

12. Calculate the osmolarity of a 0.400 M NaCl solution.
    a) 0.400   b) 0.800   c) 0.200   d) 0.100

13. What is the osmotic pressure (in torr) of a 0.0015 M NaCl solution at 25°C? R = 62.4 L torr/K mol
    a) 4.68   b) 2.34   c) 27.9   d) 55.8

True-False

14. Paint is a colloid.

15. A solution can contain only one solute.

16. A solution is a homogeneous mixture.

17. A colligative solution property is used to prevent winter freeze-up of cars.

18. Water is a good solvent for both ionic and polar covalent materials.

19. A nonpolar solute would be relatively insoluble in water.

20. On the basis of their solubility in each other, water and gasoline molecules have similar polarities.

21. Potassium nitrate is soluble in water.

22. Soluble solutes always dissolve rapidly.

Matching

Match the colloid names given as responses to the descriptions below.

23. a liquid dispersed in a gas

24. a liquid dispersed in a liquid

25. mayonnaise is an example

26. a solid dispersed in a liquid

27. gelatin dessert is an example

a) foam
b) emulsion
c) aerosol
d) sol

ANSWERS TO PROGRAMMED REVIEW

7.1 a) homogeneous   b) atoms   c) molecules   d) ions
     e) solvent   f) solutes

7.2 a) soluble   b) insoluble   c) solubility   d) saturated
     e) supersaturated

7.3 a) hydrated   b) equilibrium   c) like dissolves like   d) soluble
     e) solubility   f) solvent

7.4 a) absorbed   b) released   c) endothermic   d) cools
     e) exothermic   f) increases

7.5 a) concentrations   b) molarity   c) liter   d) percent
     e) weight/weight   f) 100   g) liquids   h) gases

7.6 a) solute   b) solvent   c) concentrated   d) solvent

7.7 a) electricity   b) electrolytes   c) non-electrolytes
     d) colligative   e) vapor   f) boiling   g) freezing
     h) semipermeable   i) osmotic

7.8 a) homogeneous   b) dispersing   c) dispersed   d) Tyndall

7.9 a) gluelike   b) state   c) liquid   d) gas

7.10 a) formation   b) destruction   c) solids   d) gaseous
     e) emulsifying   f) stabilizing

7.11 a) dialyzing   b) colloid   c) molecules   d) dialysis
     e) kidney

ANSWERS TO SELF-TEST QUESTIONS

1.	b	10.	c	19.	T
2.	a	11.	d	20.	F
3.	d	12.	b	21.	T
4.	b	13.	d	22.	F
5.	a	14.	T	23.	c
6.	b	15.	F	24.	b
7.	a	16.	T	25.	b
8.	d	17.	T	26.	d
9.	b	18.	T	27.	d

CHAPTER 8

Reaction Rates and Equilibrium

PROGRAMMED REVIEW

Section 8.1 Spontaneous and Nonspontaneous Changes

Processes that take place naturally with no apparent cause or stimulus are called (a) _____ processes. (b) _____ processes give up energy as they occur, while (c) _____ processes gain or accept energy. (d) _____ is a measurement or indication of the disorder of a system. Substances that do not undergo spontaneous changes are said to be (e) _____.

Section 8.2 Reaction Rates

The (a) _____ of a reaction is called the reaction rate. A reaction rate is determined experimentally as a change in (b) _____ of a (c) _____ or (d) _____ divided by the time required for the change. This measured rate is an (e) _____ rate for the reaction.

Section 8.3 Molecular Collisions

A detailed explanation of how a reaction takes place is called a (a) _____ _____. Reactions between particles are assumed not to take place unless the particles (b) _____ with each other. The (c) _____ _____ of molecules is the energy associated with vibrations within the molecules. The rubbing of a match head against a rough surface provides (d) _____ energy.

Section 8.4 Energy Diagrams

In an energy diagram, the vertical axis represents (a) _____, and the horizontal axis represents the (b) _____ progress. In an energy diagram, the energy of products is lower than that of reactants for an (c) _____ reaction. The energy hump on an energy diagram between the energy of reactants and products represents (d) _____ energy for the reaction.

116   CHAPTER 8

Section 8.5 Factors that Influence Reaction Rates

Four factors that influence reacction rates are the (a) _____ of the reactants, the (b) _____ of the reactants, the (c) _____ of the reactants, and the presence of (d) _____. Collisions between molecules that have the potential to cause a reaction to occur are called (e) _____ _____. (f) _____ are substances that change reaction rates without being used up in the reaction. (g) _____ _____ are ions or molecules uniformly dispersed throughout a reaction mixture, while (h) _____ _____ are used in the form of solids.

Section 8.6 Chemical Equilibrium

In principle, all reactions can occur in (a) _____ _____. When the (b) _____ and (c) _____ reaction rates are equal, the reaction is in a state of (d) _____. The concentrations of reactants and products in this state are called (e) _____ concentrations.

Section 8.7 Position of Equilibrium

The position of equilibrium for a reaction is an indication of the relative amounts of (a) _____ and (b) _____ present at equilibrium. When the equilibrium position is described as being far to the right, the concentration of (c) _____ is much higher than that of (d) _____.

Section 8.8 Factors that Influence Equilibrium Position

The influence of a number of factors on the position of equilibrium can be predicted by using (a) _____ principle. According to this principle, the addition of a reactant to an equilibrium mixture will shift the equilibrium toward the (b) _____, and heating an equilibrium mixture of an exothermic reaction will shift the equilibrium toward the (c) _____.

SOLUTIONS TO EXERCISES ANSWERED IN THE TEXT

8.1   In each case, the answer given in Appendix B of the text provides the explanation.

8.2   In each case, the answer given in Appendix B of the text provides the explanation.

Reaction Rates and Equilibrium

8.3 b) The seed needs sunlight (energy) to grow. During the growth process, the atoms and molecules of nutrients, water, etc. become organized into the tissues of the tree. This organizing constitutes an entropy decrease. Thus, the energy increases and the entropy decreases; the process is nonspontaneous.

d) Heat (energy) must leave the water in order for it to freeze. As water freezes, the somewhat random distribution of molecules characteristic of the liquid state is changed to an orderly state characteristic of solid, crystalline water. Thus, the energy decreases and the entropy decreases. From experience, we know that water does freeze spontaneously if the temperature is low enough. We conclude that the energy decrease is large enough to compensate for the entropy decrease and allow the process to take place spontaneously.

f) In order for the odor to spread, some of the liquid must evaporate. This requires an input of energy to the liquid. The random gaseous molecules have greater entropy than the more orderly liquid molecules. Thus, the energy increases and the entropy increases. We must conclude that the entropy increase is large enough to compensate for the energy increase, and allow the process to occur spontaneously.

8.4 In each case, the answer given in Appendix B of the text provides the explanation.

8.6 In each case, the answer depends upon the interpretation of the meaning of the terms *very slow, slow* and *fast*.

8.7 The answers given in Appendix B of the text are examples only. Other methods may be used and would be as good as those given.

8.8 In each case, the average rate of the reaction is calculated using the equation Rate = $\Delta C/\Delta t$, where ΔC is the change in concentration of a reactant or product that occurs in the time Δt.

a) No product, C, was present when the reaction was started, so ΔC will simply be the concentration of product at the end of the reaction. The time, Δt, is the time the reaction was allowed to take place. Thus,

$$\text{Rate} = \frac{\Delta C}{\Delta t} = \frac{(0.447 \frac{mol}{L})}{(10.0 \text{ min})}$$

or Rate = 4.47×10^{-2} mol/L/min, or 7.45×10^{-4} mol/L/s.

c) In this reaction, the concentration of reactant A started out as 0.361 M, and ended up after 7.00 minutes as 0.048 M. Thus, the value of ΔC = 0.361 M - 0.048 M = 0.313 M, and Δt = 7.00 min.

118   CHAPTER 8

$$\text{Rate} = \frac{\Delta C}{\Delta t} = \frac{(0.313 \frac{mol}{L})}{(7.00 \text{ min})} = 4.47 \times 10^{-2} \text{ mol/L/min},$$

or 7.45×10^{-4} mol/L/s

8.10   According to the ideal gas law, PV = nRT, or n = PV/RT. We will solve for n.

$$n = \frac{PV}{RT} = \frac{(1.32 \times 10^{-2} \text{ atm})(1.00 \text{ L})}{(0.0821 \frac{L \cdot atm}{K \cdot mol})(298 K)} = 5.40 \times 10^{-4} \text{ mol}$$

This gaseous product was collected in a previously empty 1-liter container, so the concentration change is equal to 5.40×10^{-4} mol/L

$$\text{Rate} = \frac{\Delta C}{\Delta t} = \frac{(5.40 \times 10^{-4} \frac{mol}{L})}{(1000 \text{ s})} = 5.40 \times 10^{-7} \text{ mol/L/s}$$

8.12   The rate of decomposition determined in Exercise 8.11 is equal to 6.0×10^{-6} mol/L/day. In this exercise, we want to find the time, Δt, that it takes for 8.0×10^{-4} mol/L of insecticide to be reduced to zero. Thus, $\Delta C = 8.0 \times 10^{-4}$ mol/L. The rate is known from exercise 8.11, so

$$\Delta t = \frac{\Delta C}{\text{Rate}} = \frac{(8.0 \times 10^{-4} \frac{mol}{L})}{(6.0 \times 10^{-6} \text{ mol/L/day})} = 133 \text{ days}$$

8.13   a) When the concentration of reactants is decreased, the number of collisions between reacting molecules would also decrease. Yet, this reaction speeds up. Thus, the reaction rate must not depend on the number of collisions between reactant molecules.
       b) If the reaction rate depended on the number of collisions between reactant molecules, the rate should increase with both an increase in A and an increase in B.

8.15   The answer in Appendix B of the text provides an explanation.

8.16   b) The described reaction would be represented by an energy diagram in which the reactants had higher energy than the products. The lack of activation energy would be represented by the lack of an energy barrier between the reactants and products. See the answer in Appendix B of the text for a sketch of the diagram.

Reaction Rates and Equilibrium 119

8.18 The primary difference between a catalyzed and uncatalyzed reaction is the height of the activation energy barrier between reactants and products. Catalysts decrease the height of the barrier. See the answer in Appendix B of the text for sketches of the diagrams.

8.19 The answer given in Appendix B of the text provides the explanations.

8.21 Both heating the reaction and increasing the concentration of reactants would increase the frequency of collisions between molecules and would increase the reaction rate. The addition of a catalyst would decrease the height of the activation energy barrier, thus increasing the rate at which reactant molecules could change into product molecules.

8.23 According to the rough rule of thumb given in the text, a decrease in reaction temperature of 10°C will cause the reaction rate to decrease by one-half, and thus double the time it takes for the reaction to be completed. In the reaction given, a drop of 10°C would decrease the temperature of the reaction from 20°C to 10° and increase the time from 3.7 hours to 7.4 hours. A second 10°C drop would take the reaction to the desired 0°C, and would once again double the time of the reaction from 7.4 hours to 14.8 or 15 hours.

8.25 The answer given in Appendix B of the text provides the explanation.

8.27 b) Since solid sugar is dissolving to form a solution, the achievement of equilibrium would be signalled when no more solid dissolved.

d) As the reaction proceeds from left to right, three moles of gas (CO and O_2) are converted into two moles of gas (CO_2). Thus, as the reaction proceeds from left to right, the decrease in moles of gas would cause a decrease in total gas pressure. The achievement of equilibrium would be signalled when the total gas pressure stopped changing.

f) In this familiar situation, the amount of money in the checking account is increased as paychecks are deposited, and decreased as checks are written to pay bills. Equilibrium at other than a value of zero in the checking account is indicated when the amount of money in the account stays constant.

8.28 In each case, the equilibrium expression will be an equation in which the equilibrium constant, K, is set equal to a fraction of concentrations raised to appropriate powers. The numerator of the fraction will consist of the product of the concentrations of materials found on the right side of the equation raised to powers equal to the coefficient of each material in the equation. The denominator of the fraction will consist of the product of the concentrations of materials found on the left side of the

120   CHAPTER 8

equation raised to powers equal to the coefficients of each material in the equation. See the answers in Appendix B of the text for the equilibrium expressions.

8.29   In each case, the equilibrium expression will be an equation like those described in exercise 8.28 above. See the answers in Appendix B of the text for the equilibrium expressions.

8.30   In each case, the materials to the left of the equation will be those in the denominator of the equilibrium expression. The coefficient preceding each material will be the exponent found on the concentration of that material in the equilibrium expression. The materials to the right of the equation will be those in the numerator of the equilibrium expression. The coefficients will be the exponent to which the concentration of each material is raised in the equilibrium expression. See the answers in Appendix B of the text for the equations.

8.32   According to the stoichiometry of the reaction, one mole of $COCl_2$ is formed for each mole of CO and Cl_2 that react. The data indicate that the number of mol/L of CO that reacted to establish equilibrium was 0.79-0.25, or 0.54 mol/L. Similarly, the number of mol/L of Cl_2 that reacted was 0.69-0.15, or 0.54 mol/L. From this, we may conclude that the number of mol/L of $COCl_2$ produced by the reaction and found in the equilibrium mixture is 0.54 mol/L.

$$K = \frac{[COCl_2]}{[CO][Cl_2]} = \frac{(0.54 \frac{mol}{L})}{(0.25 \frac{mol}{L})(0.15 \frac{mol}{L})} = 1.4 \times 10^1$$

8.34   a) The small size of K indicates that the numerator of the equilibrium expression (the concentrations of the products of the reaction) would have values much smaller than the denominator (the concentrations of the reactants). Thus, the reactants would be larger than the products.
   c) The large value of K indicates that the numerator of the equilibrium expression (the concentrations of the products of the reaction) would have values much larger than the denominator (the concentrations of the reactants). Thus, the products would be larger than the reactants.
   e) The small size of K (7.1×10^{-5}) indicates that the numerator of the equilibrium expression (the concentrations of the products of the reaction) would have values much smaller then the denominator (the concentrations of reactants). Thus, the reactants would be larger than the products.

8.35   b) Heat appears on the left of the reaction, so according to Le Chatelier's principle, the addition of heat to the system at equilibrium would shift the equilibrium toward the right in an attempt to use up the added heat.

d) The Ag⁺ ion appears on the left of the reaction, so according to Le Chatelier's principle, the removal of some Ag⁺ would shift the equilibrium to the left in an attempt to replace the removed Ag⁺.

f) Both heat and N_2 appear on the left of the reaction, so according to Le Chatelier's principle, the removal of some N_2 and some heat would both shift the equilibrium to the left in an attempt to replace the removed N_2 and the removed heat.

8.36 b) The Cl⁻ ion appears on the left of the reaction, so according to Le Chatelier's principle, the addition of some Cl⁻ to the equilibrium mixture will shift the equilibrium to the right in an attempt to use up the added Cl⁻. A shift to the right will cause the concentration of pink Co^{2+} to decrease, and the concentration of blue $CoCl_4^{2-}$ to increase. Thus, the solution becomes less pink and more blue in color.

d) Cl⁻ appears on the left of the reaction, so according to Le Chatelier's principle, the addition of some Cl⁻ will shift the equilibrium to the right in an attempt to use up the added Cl⁻. A shift to the right will produce more solid $PbCl_2$.

f) NH_3 appears on the left of the reaction, so according to Le Chatelier's principle, the removal of some NH_3 from the equilibrium mixture will shift the equilibrium to the left in an attempt to replace the removed NH_3. A shift to the left will reduce the concentration of the dark purple $Cu(NH_3)_4^{2+}$ ion and increase the concentration of the blue Cu^{2+} ion. Thus, the solution will become more blue and less purple in color.

h) Heat appears on the right of the reaction, so according to Le Chatelier's principle, the addition of heat will shift the equilibrium to the left in an attempt to use up the added heat. A shift to the left will also increase the concentration of the violet-colored I_2 gas. Thus, the mixture will become more intensely violet in color.

8.37 b) CO_2 appears on the left of the reaction, so according to Le Chatelier's principle, the removal of some CO_2 would shift the equilibrium to the left in an attempt to replace the removed CO_2. A shift to the left would increase the H_2O concentration and decrease the H_2CO_3 concentration. The CO_2 concentration was decreased by removing some of it.

d) CO_2 appears on the left of the reaction, so according to Le Chatelier's principle, the addition of some CO_2 would shift the equilibrium to the right in an attempt to use up the added CO_2. A shift to the right would decrease the LiOH concentration and increase the $LiHCO_3$ concentration. The CO_2 concentration was increased by adding it to the mixture.

f) Heat appears on the left of the equation, so according to Le Chatelier's principle, the addition of heat would shift the equilibrium to the right in an attempt to use up the added heat. A shift to the right would decrease the $CaCO_3$ concentration, and increase the CaO and CO_2 concentrations.

122 CHAPTER 8

8.38 a) H$_2$ appears on the right of the equation, so according to Le Chatelier's principle, the removal of some H$_2$ would shift the equilibrium to the right in an attempt to replace the removed H$_2$.

b) Br$_2$ appears on the right of the equation, so according to Le Chatelier's principle, the addition of some Br$_2$ would shift the equilibrium to the left in an attempt to use up the added Br$_2$.

e) HBr appears on the left of the equation, so according to Le Chatelier's principle, the addition of some HBr would shift the equilibrium to the right in an attempt to use up the added HBr.

SELF-TEST QUESTIONS

Multiple Choice

1. A catalyst
   a) is not used up in a reaction
   b) changes the rate of a reaction
   c) affects the forward reaction the same as it affects the reverse reaction
   d) more than one response is correct

2. Which of the following responses correctly arranges the states of matter for a pure substance in the order of decreasing entropy?
   a) gas, liquid, solid
   b) liquid, solid, gas
   c) solid, liquid, gas
   d) solid, gas, liquid

3. Four processes occur as the following changes take place in energy and entropy. Which process is definitely nonspontaneous?
   a) energy decrease and entropy increase
   b) energy decrease and entropy decrease
   c) energy increase and entropy increase
   d) energy increase and entropy decrease

4. A carrot cooks in 15 minutes in boiling water (100°C). How long will it take to cook a carrot inside a pressure cooker where the temperature is 10°C greater (110°C)?
   a) 3.7 minutes
   b) 5.0 minutes
   c) 7.5 minutes
   d) 30 minutes

Questions 5 and 6 refer to the following reaction, which is assumed to be at equiibrium:

$$\text{heat} + 2NO + O_2 \rightleftharpoons 2NO_2$$

In each case choose the response which best indicates the effects resulting from the described change in conditions.

5. The reaction mixture is heated.
   a) equilibrium shifts left
   b) equilibrium shifts right
   c) equilibrium does not shift
   d) the effect cannot be predicted

6. A catalyst is added to the reaction mixture.
   a) equilibrium shifts left
   b) equilibrium shifts right
   c) equilibrium does not shift
   d) the effect cannot be predicted

Question 7 refers to the following reaction:

$$2N_2O_5 \rightleftharpoons 4NO_2 + O_2$$

7. When an equilibrium constant expression is written, the exponent on the concentration of N_2O_5 is
   a) 1
   b) 2
   c) 0
   d) can't be determined from the information given

8. A sample of ICl is placed in a container and equilibrium is established according to the reaction:

   $$2ICl \rightleftharpoons I_2 + Cl_2$$

   where all the materials are gases. Analysis of the equilibrium mixture gave the following molar concentrations:

   $$[ICl] = 0.26, [I_2] = [Cl_2] = 0.09$$

   What is the value of K, the equilibrium constant for the reaction?
   a) 0.031   b) 0.12   c) 0.35   d) 14.8

True-False

9. A reaction rate can be thought of as the speed of a reaction.

10. Effective molecular collisions are those that allow molecules to collide but not react.

11. Catalysts that slow reactions are called inhibitors.

12. In a reaction at equilibrium, the forward and reverse reactions have both stopped.

13. If an endothermic reaction is spontaneous, then entropy must have decreased.

14. Catalysts act by lowering the activation energy.

15. The entropy of a cluttered room is higher than that of an orderly room.

Matching

For each of the following processes choose the appropriate response from those on the right.

16. a match burns

17. perspiration evaporates

18. melted lead becomes a solid

19. an explosive detonates

a) both entropy and energy increase
b) both entropy and energy decrease
c) entropy increases; energy decreases
d) entropy decreases; energy increases

Three liquid fuels are to be tested. A 1.0 gram sample of each fuel is weighed out and heated to its ignition temperature. When the fuel burns, the total heat liberated is measured. The results of this experiment are given in the table below. Choose the answer that best fits each statement given on the left.

Fuel	Ignition temperature	Heat liberated
X	210°C	1680 cal
Y	110°C	1410 cal
Z	285°C	1206 cal

20. it has the highest activation energy

21. it has the second highest activation energy

22. it has the lowest activation energy

23. it has the smallest energy difference between reactants and products

24. it has the largest energy difference between reactants and products

25. the described reaction is exothermic (exergonic)

a) fuel X
b) fuel Y
c) fuel Z
d) two or more fuels fit this category

Use the following equilibrium expression and match the effects on the equilibrium from the right with the changes made to the equilibrium system listed on the left.

$$\text{Heat} + CO + 2H_2 \rightleftarrows CH_3OH$$

26. add CO

27. add H_2

28. remove some CH_3OH

29. heat the system

30. add a catalyst

a) shift left
b) shift right
c) no effect on equilibrium
d) cannot be determined from the information given

ANSWERS TO PROGRAMMED REVIEW

8.1   a) spontaneous   b) exergonic   c) endergonic   d) entropy
      e) stable

8.2   a) speed   b) concentration   c) reactant   d) product
      e) average

8.3   a) reaction mechanism   b) collide   c) internal energy
      d) activation

8.4   a) energy   b) reaction   c) exergonic   d) activation

8.5   a) nature   b) concentration   c) temperature   d) catalysts
      e) effective collisions   f) catalysts   g) homogeneous catalysts
      h) heterogeneous catalysts

8.6   a) both directions   b) forward   c) reverse   d) equilibrium
      e) equilibrium

8.7   a) reactants   b) products   c) products   d) reactants

8.8   a) Le Chatelier's   b) right (or products)   c) left (or reactants)

ANSWERS TO SELF-TEST QUESTIONS

1.	d	11.	T	21.	a
2.	a	12.	F	22.	b
3.	d	13.	F	23.	c
4.	c	14.	T	24.	a
5.	b	15.	T	25.	d
6.	c	16.	c	26.	b
7.	b	17.	a	27.	b
8.	b	18.	b	28.	b
9.	T	19.	c	29.	b
10.	F	20.	c	30.	c

CHAPTER 9

Acids, Bases and Salts

PROGRAMMED REVIEW

Section 9.1 The Arrhenius Theory

Svante Arrhenius defined an acid as a substance that (a) _____ when dissolved in water and produces (b) _____ ions. He defined a base as a substance that (c) _____ and releases (d) _____ ions when dissolved in water.

Section 9.2 The Brønsted Theory

According to the Brønsted theory an acid is any (a) _____ containing substance that is capable of donating a (b) _____ to another substance. A base is any substance that accepts a (c) _____. The species that remains when a Brønsted acid donates a proton is called the (d) _____ base of the acid.

Section 9.3 The Self Ionization of water

In the pure state, water undergoes a self or (a) _____ ionization in which some water molecules function as Bronsted (b) _____ and some function as Bronsted (c) _____. The term *neutral* is used to describe any water solution that contains equal concentrations of (d) _____ and (e) _____ ions. A solution is acidic when the (f) _____ ion concentration is higher, and basic or alkaline when the (g) _____ ion concentration is higher.

Section 9.4 Properties of Acids

All acids taste (a) _____ and produce (b) _____ ions when dissolved in water. In addition, all acids react characteristically with solid (c) _____, (d) _____, (e) _____ and (f) _____. Acids react with certain metals to produce (g) _____ gas. The tendency of metals to undergo such a reaction is given by an (h) _____ _____.

127

Section 9.5 Properties of Bases

The complete reaction of an (a) _____ with a (b) _____ to produce a solution containing only (c) _____ and a (d) _____ is called a (e) _____ reaction. Bases are often found in household cleaners because they react with (f) _____ and (g) _____.

Section 9.6 Salts

At room temperature, salts are solid (a) _____ substances that contain the (b) _____ of a base and the (c) _____ of an acid. Salts that contain specific numbers of water molecules as a part of their crystalline structure are called (d) _____. The water in such compounds is called the (e) _____ _____ _____. An equivalent of a salt is the amount of salt that will produce one mole of (f) _____ _____ when dissolved and dissociated.

Section 9.7 The pH Concept

Mathematically, pH is the (a) _____ _____ of the (b) _____ concentration of the (c) _____ ion in a solution. A solution with a pH higher than 7 is classified as (d) _____, and one lower than 7 is classified as (e) _____.

Section 9.8 Strength of Acids and Bases

The strength of an acid or base is determined by the extent to which they (a) _____ when dissolved in water. Strong acids or bases dissociate essentially (b) _____ in solution. Acids are classified as monoprotic, diprotic and triprotic on the basis of the number of (c) _____ given up per molecule. In general, the anions produced by the dissociation of strong Brønsted acids behave as (d) _____ Brønsted bases.

Section 9.9 Analysis of Acids and Bases

A common procedure used to analyze acids and bases is called (a) _____. When this procedure is used to analyze an acid, a (b) _____ solution of known concentration is added to an acid solution until the (c) _____ point is reached. One way to detect when an acid and base have reacted completely, is to add an (d) _____ which changes color as close as possible to the (e) _____ point of the titration.

Section 9.10 Titration Calculations

Titration calculations are done using the methods described earlier in Chapter (a) _____ of the text.

Section 9.11 Hydrolysis Reactions of Salts

In general, a hydrolysis reaction is any reaction with (a) _____. In the case of salts, hydrolysis reactions cause salt solutions to have a (b) _____ different from that of pure (c) _____. The hydrolysis of a salt containing the cation of a strong base and the anion of a weak acid produces a (d) _____ solution.

Section 9.12 Buffers

Buffers are solutions with the ability to resist changes in (a) _____ when (b) _____ or (c) _____ are added. The amount of H^+ or OH^- that a buffer system can absorb without changing (d) _____ significantly is called the (e) _____ _____.

SOLUTIONS TO EXERCISES ANSWERED IN THE TEXT

9.1 In each case, H^+ should be produced to emphasize the Arrhenius acid characteristic. Also note that the charges of the ions must balance on each side of the equation.
   b) $HBrO \rightarrow H^+ + BrO^-$
   d) $HClO_2 \rightarrow H^+ + ClO_2^-$
   f) $H_3PO_2 \rightarrow H^+ + H_2PO_2^-$

9.2 In order to behave as an Arrhenius base, the compound must contain OH^- ions.
   a) CsOH contains the OH^- ion, and will behave as an Arrhenius base in water.
      $CsOH \rightarrow Cs^+ + OH^-$
   c) NH_3 does not contain OH^-, and so cannot behave as an Arrhenius base.
   e) Na_2S does not contain OH^-, and so cannot behave as an Arrhenius base.

9.3 The answers given in Appendix B of the text provide the explanations.

9.4 In each case, a Brønsted acid will be any species that gives up a proton, H^+, and a Brønsted base will be any species that accepts a proton, H^+. Because all the reactions are reversible, each reaction will contain two Brønsted acids and two Brønsted bases. A species that accepts a proton in the forward reaction (it behaves as a base) is transformed into a species that will give up a proton in the reverse reaction (it behaves as an acid).
   b) Acids: H_2O (forward reaction), HN_3 (reverse reaction)
      Bases: N_3^- (forward reaction), OH^- (reverse reaction)
   d) Acids: H_2O (forward reaction), HSO_3^- (reverse reaction)

Bases: SO$_3^{2-}$ (forward reaction), OH⁻ (reverse reaction)
f) Acids: H$_2$O (forward reaction), CH$_3$NH$_2$ (reverse reaction)
Bases: CH$_3$NH⁻ (forward reaction), OH⁻ (reverse reaction)
h) Acids: HN$_3$ (forward reaction), H$_3$O⁺ (reverse reaction)
Bases: H$_2$O (forward reaction), N$_3^-$ (reverse reaction)

9.5 In each case, the species that accepts a proton in the forward reaction (it acts as a base) is transformed into an acid for the reverse reaction. The two are called a conjugate pair. The species that gives up the proton in the forward reaction (it behaves as an acid) is transformed into a base for the reverse reaction. The two are called a conjugate pair. See the answers in Appendix B of the text for diagrams that illustrate this point.

9.6 In each case, the conjugate base is formed by removing one H⁺ ion from the acid. Note that the conjugate base will have one more negative charge than the acid from which it was formed. In the following reactions, the conjugate base asked for in the exercise is underlined.
a) HSO$_3^-$ → H⁺ + $\underline{SO_3^{2-}}$
c) HClO$_3$ → H⁺ + $\underline{ClO_3^-}$
e) H$_2$C$_2$O$_4$ → H⁺ + $\underline{HC_2O_4^-}$
g) HBrO$_3$ → H⁺ + $\underline{BrO_3^-}$
i) HCO$_3^-$ → H⁺ + $\underline{CO_3^{2-}}$

9.8 In each case, the conjugate acid is formed by adding one H⁺ ion to the species acting as the base. Note that the conjugate acid will have one less negative charge than the base from which it was formed. In the following reactions, the conjugate acid asked for in the exercise is underlined.
a) NH$_2^-$ + H⁺ → $\underline{NH_3}$
c) OH⁻ + H⁺ → $\underline{H_2O}$
e) NO$_2^-$ + H⁺ → $\underline{HNO_2}$
g) HPO$_4^{2-}$ + H⁺ → $\underline{H_2PO_4^-}$
i) I⁻ + H⁺ → $\underline{HI}$

9.9 b) The needed species forms NH$_4^+$ when it accepts a H⁺ ion. The species is an NH$_3$ molecule.
d) The needed species is formed when a H⁺ ion is removed from H$_2$N$_2$O$_2$. The species is HN$_2$O$_2^-$.
f) The needed species is one that forms N$_2$H$_5^+$ when it accepts a H⁺ from water. The species is N$_2$H$_4$.
h) The needed species is formed when H$_3$AsO$_3$ gives up a H⁺. The species is H$_2$AsO$_3^-$.

9.10 In each case, the acid will give up a H⁺ and the base will accept it. The acid will be transformed into a conjugate base with one more negative charge than the acid had originally. The base will be transformed into a conjugate acid with one more positive charge than the base had originally. See the answers in Appendix B of the text for the equations.

Acids, Bases and Salts   131

9.11 In each case the following equation (Equation 9.10) will be used in a rearranged form. $1.0 \times 10^{-14} = [H_3O^+][OH^-]$. This can be rearranged to solve for $[OH^-]$ as follows:

$$[OH^-] = \frac{(1.0 \times 10^{-14})}{[H_3O^+]}$$

This rearranged form of the equation will be used for each calculation.

b) $[OH^-] = \dfrac{(1.0 \times 10^{-14})}{(3.2 \times 10^{-3})} = 3.1 \times 10^{-12}$

d) $[OH^-] = \dfrac{(1.0 \times 10^{-14})}{(7.7 \times 10^{-9})} = 1.3 \times 10^{-6}$

f) $[OH^-] = \dfrac{(1.0 \times 10^{-14})}{0.043} = 2.3 \times 10^{-13}$

h) $[OH^-] = \dfrac{(1.0 \times 10^{-14})}{(1.0 \times 10^{-8})} = 1.0 \times 10^{-6}$

9.12 Equation 9.10 used in exercise 9.11 can be rearranged to solve for $[H_3O^+]$ as follows:

$$[H_3O^+] = \frac{(1.0 \times 10^{-14})}{[OH^-]}$$

This rearranged form of the equation will be used for each calculation.

b) $[H_3O^+] = \dfrac{(1.0 \times 10^{-14})}{(5.2 \times 10^{-4})} = 1.9 \times 10^{-11}$

d) $[H_3O^+] = \dfrac{(1.0 \times 10^{-14})}{(3.8 \times 10^{-12})} = 2.6 \times 10^{-3}$

f) $[H_3O^+] = \dfrac{(1.0 \times 10^{-14})}{(3.7)} = 2.7 \times 10^{-15}$

CHAPTER 9

h) $[H_3O^+] = \dfrac{(1.0 \times 20^{-14})}{(0.00044)} = 2.3 \times 10^{-11}$

9.13 In exercise 9.11, any solution with [OH⁻] greater than 1.0×10^{-7} will be classified as basic. If the [OH⁻] is less than 1.0×10^{-7}, the solution is acidic.
b) acidic   d) basic   f) acidic   h) basic
In exercise 9.12, any solution with $[H_3O^+]$ greater than 1.0×10^{-7} will be classified as acidic. If the $[H_3O^+]$ is less than 1.0×10^{-7}, the solution is basic.
b) basic   d) acidic   f) basic   h) basic

9.14 The answers given in Appendix B of the text provide the explanations.

9.15 In each case Equation 7.6 is used to calculate the volume of the concentrated solution needed to prepare the dilute solution.
b) Stock sodium hydroxide solution is 6 M. Equation 7.6 is rearranged to solve for V_c, the volume of the stock solution needed:

$$V_c = \dfrac{V_d C_d}{C_c} = \dfrac{(200\ mL)(0.1\ M)}{(6\ M)} = 3.3\ mL$$

To prepare the 0.1 M solution, put 3.3 mL of 6 M NaOH solution into a container and add 197 mL of distilled water.

d) Dilute hydrochloric acid solution is 6 M. Equation 7.6 is rearranged to solve for V_c, the volume of 6 M HCl needed:

$$V_c = \dfrac{V_d C_d}{C_c} = \dfrac{(5\ L)(1.5\ M)}{(6\ M)} = 1.3\ L$$

To prepare the solution, put 1.3 L of dilute HCl solution into a 5-L container, add 3.7 L of distilled water, and stir well.

f) Concentrated aqueous ammonia is 15 M. Equation 7.6 is rearranged to solve for V_c, the volume of 15 M NH_3 solution needed:

$$V_c = \dfrac{V_d C_d}{C_c} = \dfrac{(2.5\ L)(1.0\ M)}{(15\ M)} = 0.167\ L = 167\ mL$$

To prepare the solution, put 167 mL of 15 M NH_3 into a container and add enough distilled water to give a final volume of 2.5 L. Stir well.

Acids, Bases and Salts   133

9.16   In each case, remember that a full equation contains the complete formulas of all reactants and products. The equations are given as answers in Appendix B of the text.

9.17   In each case, remember to write soluble materials in the form of the ions they contain. Insoluble materials are written using their complete formulas. The net ionic equation is obtained by eliminating any ions that appear on both sides of the total ionic equation. The equations are given as answers in Appendix B of the text.

9.19   In each case, one of the products will be hydrogen gas, H_2. The other products will be the metal ion and the anion characteristic of the acid. In the full equation, the metal ion and anion of the acid will be written as a salt. Since each metal forms a 2+ ion, the stoichiometry in each reaction will require two moles of H^+ to react with one mole of metal. In those cases where the acid contains a single hydrogen, two moles of acid will be required. In those cases where the acid contains two hydrogens, one mole of acid will be required. The equations are given as answers in Appendix B of the text.

9.20   In each case, one H^+ from the acid will react with one LiOH. Thus, the coefficient on LiOH in the balanced equation will be the same as the number of reacting H's in the acid. Remember that the reacting H's of an acid are those written first in the formula of the acid. The equations are given as answers in Appendix B of the text.

9.21   In each case, one H^+ from the acid will react with one KOH. Thus, the coefficient on KOH in the balanced equation will be the same as the number of H's reacting per acid molecule. The anion of the acid left to form the salt might contain H's, depending upon the number of H's reacted per molecule of acid. The charge on the acid anion will be negative, and equal in number to the number of H's reacted per molecule of acid. The equations are given in Appendix B of the text.

9.22   With the exception of the ammonium ion, NH_4^+, the cation in each salt is a metallic cation. Reference to Table 4.6 will provide the charges of the various polyatomic anions involved. The assignment of charges to the cations can be done by balancing the total positive charge of the cations in the salt with the total negative charge of the anions in the salt. The cations and anions are listed as answers in Appendix B of the text.

9.23   In each case, the formula of the acid involved will be the anion of the salt with enough H's added to match the negative charge of the anion. For example, if the anion were $H_2PO_4^-$, the acid would be H_3PO_4, where one H^+ has been added to the anion to balance the -1 charge of the anion. The base in each case can be obtained by adding enough OH^- ions to the cation to balance the cation charge. For example if the cation were Mg^{2+}, the formula of the base would be $Mg(OH)_2$, where two OH^-

134   CHAPTER 9

ions have been added to the cation to balance the +2 charge of the cation. The formulas of the acids and bases are given as answers in Appendix B of the text.

9.24 a) The formula for plaster of paris is $CaSO_4 \cdot H_2O$. Thus, we see that each mole of plaster of paris contains one mole of water. The mass of water released when the water of hydration of one mole of plaster of paris is driven off would be the mass of one mole of water or 18 g.

9.25 In each case, the anion portion of the compound will come from the acid used, and the cation portion will come from the indicated solid.
b) $MgSO_4$: The anion comes from sulfuric acid, H_2SO_4 (see Table 9.1 for acid formulas). The cation comes from a carbonate of magnesium. According to Table 4.6, the formula for the carbonate polyatomic ion is CO_3^{2-}. The Mg^{2+} and CO_3^{2-} ions combine in a 1:1 ratio, so the solid used is $MgCO_3$.
d) $Zn(NO_3)_2$: The anion comes from nitric acid, HNO_3 (see Table 9.1). The cation comes from a metal, so the metal must be zinc metal, Zn.
f) KI: The anion comes from the acid HI, which is not given in Table 9.1, but which is analogous to HCl. The cation comes from an oxide. The oxide ion, O^{2-}, has a charge of -2 (oxygen is in group VI A (16) of the periodic table), and the potassium ion, K^+, has a charge of +1 (potassium is in group I A (1) of the periodic table). These two ions will combine in a 1:2 ratio respectively, so the formula for the oxide is K_2O.

9.26 Each reaction is a characteristic reaction of acids. The reactions are given in Appendix B of the text.

9.27 One equivalent of a salt is the amount of salt (number of moles) that would produce one mole of positive (or negative) charges when dissolved and dissociated.
b) Li_2CO_3: The dissociation reaction is: $Li_2CO_3 \rightarrow 2\ Li^+ + CO_3^{2-}$.
We see that when one mole of Li_2CO_3 dissolves and dissociates, two moles of Li^+ (positive charges) is produced. Thus, one-half mole of Li_2CO_3 would produce one mole of positive charges and would thus be equal to one equivalent.
d) $Zn(NO_3)_2$: The dissociation reaction is: $Zn(NO_3)_2 \rightarrow Zn^{2+} + 2\ NO_3^-$. We see that when one mole of $Zn(NO_3)_2$ dissolves and dissociates, one mole of Zn^{2+} is produced. However, since each ion carries two positive charges, this is two moles of positive charges. Thus, one-half mole of $Zn(NO_3)_2$ would produce one mole of positive charges and would be equal to one equivalent.
f) NaI: The dissociation reaction is: $NaI \rightarrow Na^+ + I^-$. We see that when one mole of NaI dissolves and dissociates, one mole of Na^+ is produced. This is also one mole of positive charges, so we conclude that one mole of NaI is equal to one equivalent.

Acids, Bases and Salts 135

9.28 b) For MgCl$_2$, the dissociation reaction is: MgCl$_2$ → Mg^{2+} + 2 Cl$^-$. Thus, we see that one mole of MgCl$_2$ gives two moles of positive charges, or 1 mol = 2 eq. The number of equivalents can be calculated using this relationship.

$$0.25 \text{ mol MgCl}_2 \times \frac{2 \text{ eq MgCl}_2}{1 \text{ mol MgCl}_2} = 0.50 \text{ eq MgCl}_2$$

$$0.50 \text{ eq MgCl}_2 \times \frac{1000 \text{ meq}}{1 \text{ eq}} = 5.0 \times 10^2 \text{ meq MgCl}_2$$

d) For FeSO$_4$, the dissociation reaction is: FeSO$_4$ → Fe^{2+} + SO$_4^{2-}$. Thus, we see that one mole of the salt gives one mole of Fe^{2+} or two moles of positive charges, or 1 mol = 2 eq. The number of equivalents can be calculated using this relationship.

$$0.50 \text{ mol FeSO}_4 \times \frac{2 \text{ eq FeSO}_4}{1 \text{ mol FeSO}_4} = 1.0 \text{ eq FeSO}_4$$

$$1.0 \text{ eq FeSO}_4 \times \frac{1000 \text{ meq}}{1 \text{ eq}} = 1.0 \times 10^3 \text{ meq FeSO}_4$$

f) For AgNO$_3$, the dissociation reaction is: AgNO$_3$ → Ag$^+$ + NO$_3^-$. Thus, we see that one mole of the salt gives one mole of Ag$^+$ or one mole of positive charges, or 1 mol = 1 eq. The number of equivalents can be calculated using this relationship.

$$4.75 \times 10^{-2} \text{ mol AgNO}_3 \times \frac{1 \text{ eq AgNO}_3}{1 \text{ mol AgNO}_3} = 4.75 \times 10^{-2} \text{ eq AgNO}_3$$

$$4.75 \times 10^{-2} \text{ eq AgNO}_3 \times \frac{1000 \text{ meq}}{1 \text{ eq}} = 4.75 \times 10^1 \text{ meq AgNO}_3$$

9.29 In each case it will be necessary to calculate the number of moles of salt in 5 grams. The conversion factor for each calculation comes from the formula weight of the salt.
b) For NaNO$_3$, the dissociation reaction is: NaNO$_3$ → Na$^+$ + NO$_3^-$. Thus, we see that one mole of salt gives one mole of Na$^+$ or one mole of positive charges, or 1 mol = 1 eq. This fact will be combined with the calculation of moles below to give equivalents.

136 CHAPTER 9

$$5.00 \text{ g NaNO}_3 \times \frac{1 \text{ mol NaNO}_3}{85.0 \text{ g NaNO}_3} \times \frac{1 \text{ eq}}{1 \text{ mol}} = 5.88 \times 10^{-2} \text{ eq}$$

$$5.88 \times 10^{-2} \text{ eq} \times \frac{1000 \text{ meq}}{1 \text{ eq}} = 5.88 \times 10^{1} \text{ meq}$$

d) For MgSO$_4$, the dissociation reaction is: MgSO$_4 \rightarrow$ Mg^{2+} + SO$_4^{2-}$. Thus, we see that one mole of salt gives one mole of Mg^{2+} or two moles of positive charges, or 1 mol = 2 eq. This fact will be used in the following calculation.

$$5.00 \text{ g MgSO}_4 \cdot 7H_2O \times \frac{1 \text{ mol MgSO}_4 \cdot 7H_2O}{246.5 \text{ g MgSO}_4 \cdot 7H_2O} \times \frac{2 \text{ eq}}{1 \text{ mol}} = 4.06 \times 10^{-2} \text{ eq}$$

$$4.06 \times 10^{-2} \text{ eq} \times \frac{1000 \text{ meq}}{1 \text{ eq}} = 4.06 \times 10^{1} \text{ meq}$$

f) For CuSO$_4$, the dissociation reaction is: CuSO$_4 \rightarrow$ Cu^{2+} + SO$_4^{2-}$. Thus, we see that one mole of salt gives one mole of Cu^{2+} or two moles of positive charges, or 1 mol = 2 eq. This fact will be used in the following calculation.

$$5.00 \text{ g CuSO}_4 \cdot 5H_2O \times \frac{1 \text{ mol CuSO}_4 \cdot 5H_2O}{249.7 \text{ g CuSO}_4 \cdot 5H_2O} \times \frac{2 \text{ eq}}{1 \text{ mol}} = 4.00 \times 10^{-2} \text{ eq}$$

$$4.00 \times 10^{-2} \text{ eq} \times \frac{1000 \text{ meq}}{1 \text{ eq}} = 4.00 \times 10^{1} \text{ meq}$$

h) For SrCl$_2$, the dissociation reaction is: SrCl$_2 \rightarrow$ Sr^{2+} + 2Cl$^-$. Thus, we see that one mole of salt gives one mole of Sr^{2+} or two moles of positive charges, or 1 mol = 2 eq. This fact will be used in the following calculation.

$$5.00 \text{ g SrCl}_2 \times \frac{1 \text{ mol SrCl}_2}{158.5 \text{ g SrCl}_2} \times \frac{2 \text{ eq}}{1 \text{ mol}} = 6.31 \times 10^{-2} \text{ eq}$$

Acids, Bases and Salts   137

$$6.31 \times 10^{-2} \; \cancel{eq} \times \frac{1000 \; meq}{1 \; \cancel{eq}} = 6.31 \times 10^{1} \; meq$$

9.30   In each case, a pH greater than 7 will be classified basic, a pH less than 7 will be classified acidic, and a pH equal to 7 will be classified neutral. See the answers in Appendix B of the text for the classifications.

9.31   In each case, pH will be calculated as pH = -log [H⁺]. In those cases where the molar concentration of OH⁻ is given rather than the molarity of H⁺, the value of [H⁺] will be calculated using the following equation:

$$[H^+] = \frac{1.0 \times 10^{-14}}{[OH^-]}$$

b) $[H^+] = \dfrac{1.0 \times 10^{-14}}{6.0 \times 10^{-3}} = 1.7 \times 10^{-12} \; M$

pH = -log(1.7 x 10⁻¹²) = 11.77. The solution is basic.

d) pH = -log(9.0 x 10⁻⁴) = 3.05. The solution is acidic.

f) $[H^+] = \dfrac{1.0 \times 10^{-14}}{7.0 \times 10^{-12}} = 1.4 \times 10^{-3} \; M$

pH = -log(1.4 x 10⁻³) = 2.85. The solution is acidic.

h) $[H^+] = \dfrac{1.0 \times 10^{-14}}{2.0 \times 10^{-5}} = 5.0 \times 10^{-10} \; M$

pH = -log(5.0 x 10⁻¹⁰) = 9.30. The solution is basic.

9.32   These problems will be solved using the same technique that was used in exercise 9.31 above.
b) pH = -log(7.4 x 10⁻⁸) = 7.13. The solution is basic.
d) pH = -log(4.4 x 10⁻⁶) = 5.36. The solution is acidic.

f) $[H^+] = \dfrac{1.0 \times 10^{-14}}{9.7 \times 10^{-12}} = 1.0 \times 10^{-3} \; M$

pH = -log(1.0 x 10⁻³) = 3.00. The solution is acidic.

138 CHAPTER 9

h) $[H^+] = \dfrac{1.0 \times 10^{-14}}{1.3 \times 10^{-1}} = 7.7 \times 10^{-14} \ M$

$pH = -\log(7.7 \times 10^{-14}) = 13.11$. The solution is basic.

9.33 In each case, the value of [H⁺] will be calculated using the following equation.

$$[H^+] = \text{antilog}(-pH)$$

b) $[H^+] = \text{antilog}(-13.12) = 7.6 \times 10^{-14} \ M$
d) $[H^+] = \text{antilog}(-9.01) = 9.8 \times 10^{-10} \ M$
f) $[H^+] = \text{antilog}(-2.17) = 6.8 \times 10^{-3} \ M$

9.34 In each case, the [H⁺] value will be determined the same way it was in exercise 9.33 above. The [OH⁻] value will then be determined using the following equation.

$$[OH^-] = \dfrac{1.0 \times 10^{-14}}{[H^+]}$$

b) $[H^+] = \text{antilog}(-6.15) = 7.1 \times 10^{-7} \ M$

$[OH^-] = \dfrac{1.0 \times 10^{-14}}{7.1 \times 10^{-7}} = 1.4 \times 10^{-8} \ M$

d) $[H^+] = \text{antilog}(-9.00) = 1.0 \times 10^{-9} \ M$

$[OH^-] = \dfrac{1.0 \times 10^{-14}}{1.0 \times 10^{-9}} = 1.0 \times 10^{-5} \ M$

f) $[H^+] = \text{antilog}(-1.52) = 3.0 \times 10^{-2} \ M$

$[OH^-] = \dfrac{1.0 \times 10^{-14}}{3.0 \times 10^{-2}} = 3.3 \times 10^{-13} \ M$

9.35 In each case, the following equation will be used:

$$[H^+] = \text{antilog}(-pH).$$

b) $[H^+] = \text{antilog}(-1.60) = 2.5 \times 10^{-2} \ M$. The solution is acidic.
d) $[H^+] = \text{antilog}(-6.85) = 1.4 \times 10^{-7} \ M$. The solution is acidic.
f) $[H^+] = \text{antilog}(-8.05) = 8.9 \times 10^{-9} \ M$. The solution is basic.

Acids, Bases and Salts 139

9.36 In each case, the following equation will be used:

$$[H^+] = \text{antilog}\,(-pH).$$

b) $[H^+]$ = antilog (-5.10) = 7.9 x 10^{-6} M. The solution is acidic.
d) $[H^+]$ = antilog (-2.65) = 2.2 x 10^{-3} M. The solution is acidic.
f) $[H^+]$ = antilog (-4.11) = 7.8 x 10^{-5} M. The solution is acidic.

9.38 The strength of the acids is indicated by the value of K_a, with the stronger acid having a larger K_a value. Thus, the acids arranged from weakest to strongest would be acid B, acid A, acid C, and acid D.

9.40 Acid strength is indicated by the value of K_a, with the acid strength increasing as the value of K_a increases.
a) On the basis of the above statement, the acids arranged with the weakest first and the strongest last are: acid B, acid A, acid C, and acid D.
b) The strength of the conjugate base of an acid depends on the strength of the acid from which it formed. The stronger an acid, the weaker it's conjugate base. Thus, when the conjugate bases are arranged in order from weakest to strongest, the order will be the reverse of that for the acids that produced the bases: base D, base C, base A, and base B.

9.41 The dissociation reactions will follow the pattern demonstrated in general by Equation 9.31, and for a specific acid by Equation 9.35. The K_a expressions will follow the pattern demonstrated in general by Equation 9.34, and for some specific acids in example 9.9. See the answers in Appendix B of the text for the reactions and expressions.

9.43 See the answer in Appendix B of the text for the explanation.

9.45 Titration is an analytical technique that allows the amount of acid (if base is used to titrate an acid) or base (if acid is used to titrate a base) contained in the sample that is titrated to be determined.

9.47 See the answer in Appendix B of the text for the explanation.

9.48 According to Figure 7.8, this problem has the pattern L soln. A → mol B, and follows the pathway L HNO_3 soln. → mol HNO_3 → mol NaOH. The necessary factors will come from the concentration of HNO_3 solution and the coefficients in the reaction equation. The reaction equation is: $HNO_3 + NaOH \rightarrow H_2O + NaNO_3$

140   CHAPTER 9

$$0.500 \text{ L HNO}_3 \text{ soln} \times \frac{0.150 \text{ mol HNO}_3}{1 \text{ L HNO}_3 \text{ soln}} \times \frac{1 \text{ mol NaOH}}{1 \text{ mol HNO}_3} = 0.0750 \text{ mol NaOH}$$

9.49   In each case, the coefficient of NaOH in the equation will be equal to the number of H's on the acid being titrated. The salt produced will have as many Na's as the original acid had H's. See the answers in Appendix B of the text for the equations.

9.50   In each case, the coefficient of HCl in the equation will be equal to the number of OH's in the formula of the base. The formula of the resulting salts will have as many Cl's as there were OH's in the original base. See the answers in Appendix B of the text for the equations.

9.52   The volume of the H_2SO_4 solution titrated is known, so the molarity could be calculated if we knew the number of moles of H_2SO_4 in the titrated sample. The calculation of the number of moles is a problem with the pattern L soln. A → mol B, with the pathway L NaOH soln. → mol NaOH → mol H_2SO_4. The necessary factors will come from the concentration of the NaOH solution, and the coefficients of the reaction equation. The reaction is $H_2SO_4 + 2NaOH \rightarrow 2H_2O + Na_2SO_4$

$$0.0292 \text{ L NaOH soln} \times \frac{0.50 \text{ mol NaOH}}{1 \text{ L NaOH soln}} \times \frac{1 \text{ mol } H_2SO_4}{2 \text{ mol NaOH}} = 0.0073 \text{ mol } H_2SO_4$$

The molarity of the H_2SO_4 solution is given by the following equation.

$$M = \frac{(\text{mol } H_2SO_4)}{(L \text{ soln})} = \frac{(0.0073 \text{ mol})}{(0.0250 \text{ L})} = 0.29 \text{ M}$$

9.53   Each of these problems has the pattern L soln A → L soln B, with the pathway L acid soln → mol acid → mol NaOH → L NaOH soln. The necessary factors will come from the solution concentrations and the coefficients in the reaction equation.
b) The reaction equation is: $H_2SO_4 + 2NaOH \rightarrow 2H_2O + Na_2SO_4$

$$0.02000 \text{ L } H_2SO_4 \text{ soln} \times \frac{0.180 \text{ mol } H_2SO_4}{1 \text{ L } H_2SO_4 \text{ soln}} \times \frac{2 \text{ mol NaOH}}{1 \text{ mol } H_2SO_4}$$

$$\times \frac{1 \text{ L NaOH soln}}{0.120 \text{ mol NaOH}} = 0.0600 \text{ L, or } 600 \text{ mL}$$

d) The reaction equation is the same as in part b above.

$$0.02000 \text{ L } H_2SO_4 \text{ soln} \times \frac{0.100 \text{ mol } H_2SO_4}{1 \text{ L } H_2SO_4 \text{ soln}} \times \frac{2 \text{ mol NaOH}}{1 \text{ mol } H_2SO_4}$$

$$\times \frac{1 \text{ L NaOH soln}}{0.120 \text{ mol NaOH}} = 0.0333 \text{ L, or } 333 \text{ mL}$$

f) The molarity of the acid solution must first be calculated. This means the number of moles of acid in 10.00 grams must be determined:

$$10.00 \text{ g } H_3PO_4 \times \frac{1 \text{ mol } H_3PO_4}{97.99 \text{ g } H_3PO_4} = 0.1021 \text{ moles}$$

The solution molarity is calculated as:

$$M = \frac{(\text{mol acid})}{(\text{L soln})} = \frac{(0.1021 \text{ mol})}{(0.250 \text{ L soln})} = 0.408 \text{ M}$$

The reaction equation is: $H_3PO_4 + 3NaOH \rightarrow 3H_2O + Na_3PO_4$

$$0.02000 \text{ L } H_3PO_4 \text{ soln} \times \frac{0.408 \text{ mol } H_3PO_4}{1 \text{ L } H_3PO_4 \text{ soln}} \times \frac{3 \text{ mol NaOH}}{1 \text{ mol } H_3PO_4}$$

$$\times \frac{1 \text{ L NaOH soln}}{0.120 \text{ mol NaOH}} = 0.204 \text{ L, or } 204 \text{ mL}$$

h) The molarity of the acid solution is first calculated as:

$$M = \frac{(\text{mol acid})}{(\text{L soln})} = \frac{(0.215 \text{ mol})}{(0.700 \text{ L soln})} = 0.307 \text{ M}$$

The reaction equation is $H_2MoO_4 + 2NaOH \rightarrow 2H_2O + Na_2MoO_4$

$$0.02000 \cancel{L\, H_2MoO_4\, soln} \times \frac{0.307 \cancel{mol\, H_2MoO_4}}{1 \cancel{L\, H_2MoO_4\, soln}} \times \frac{2 \cancel{mol\, NaOH}}{1 \cancel{mol\, H_2MoO_4}}$$

$$\times \frac{1\, L\, NaOH\, soln}{0.120 \cancel{mol\, NaOH}} = 0.102\, L,\, or\, 102\, mL$$

9.54 In each case the number of moles of acid titrated will be determined, then divided by the acid sample volume in liters. The calculation of the number of moles of acid has the pattern L soln A → mol B, and follows the pathway L base soln → mol base → mol acid. The necessary factors come from the concentration of the base, and the coefficients of the reaction equation.

b) The reaction equation is: $H_2SO_4 + 2KOH \rightarrow 2H_2O + K_2SO_4$

$$0.01112 \cancel{L\, KOH\, soln} \times \frac{0.109 \cancel{mol\, KOH}}{1 \cancel{L\, KOH\, soln}} \times \frac{1\, mol\, H_2SO_4}{2 \cancel{mol\, KOH}}$$

$$= 6.06 \times 10^{-4}\, mol\, H_2SO_4$$

$$M = \frac{(mol\, acid)}{(L\, soln)} = \frac{(6.06 \times 10^{-4}\, mol\, H_2SO_4)}{(0.02000\, L\, soln)} = 0.0303\, M$$

d) The reaction equation is: $H_2SO_4 + 2NaOH \rightarrow 2H_2O + Na_2SO_4$.

$$0.03217 \cancel{L\, NaOH\, soln} \times \frac{1.09 \cancel{mol\, NaOH}}{1 \cancel{L\, NaOH\, soln}} \times \frac{1\, mol\, H_2SO_4}{2 \cancel{mol\, NaOH}}$$

$$= 0.01753\, mol\, H_2SO_4$$

$$M = \frac{(mol\, acid)}{(L\, soln)} = \frac{(0.01753\, mol\, H_2SO_4)}{(0.005\, L\, soln)} = 3.51\, M$$

f) The reaction equation is: $HCl + NaOH \rightarrow H_2O + NaCl$

$$0.04125 \text{ L NaOH soln} \times \frac{3.00 \text{ mol NaOH}}{1 \text{ L NaOH soln}} \times \frac{1 \text{ mol HCl}}{1 \text{ mol NaOH}}$$

$$= 0.124 \text{ mol HCl}$$

$$M = \frac{(\text{mol acid})}{(\text{L soln})} = \frac{(0.124 \text{ mol HCl})}{(0.01000 \text{ L soln})} = 12.4 \text{ M}$$

9.55 The reaction equation is: $H_2C_2O_4 + 2NaOH \rightarrow 2H_2O + Na_2C_2O_4$
This problem has the pattern L soln. A → mol B, with the pathway L NaOH soln. → mol NaOH → mol $H_2C_2O_4$. The necessary factors will come from the concentration of the NaOH solution and the coefficients in the reaction equation.

$$0.02786 \text{ L NaOH soln} \times \frac{0.250 \text{ mol NaOH}}{1 \text{ L NaOH soln}} \times \frac{1 \text{ mol } H_2C_2O_4}{2 \text{ mol NaOH}}$$

$$= 3.48 \times 10^{-3} \text{ mol } H_2C_2O_4$$

9.56 The reaction equation is $HBz + NaOH \rightarrow H_2O + NaBz$, where HBz represents the monoprotic benzoic acid. The molecular weight can be determined by dividing the mass of a sample in grams by the number of moles in the sample. The mass of a titrated sample is given, so the titration data are used to calculate the number of moles of acid in the titrated sample. This problem has the pattern L soln A → mol B, with the pathway L NaOH soln. → mol NaOH → mol HBz. The necessary factors will come from the NaOH solution concentration, and the coefficients of the reaction equation.

$$0.04675 \text{ L NaOH soln} \times \frac{0.1021 \text{ mol NaOH}}{1 \text{ L NaOH soln}} \times \frac{1 \text{ mol HBz}}{1 \text{ mol NaOH}}$$

$$= 4.77 \times 10^{-3} \text{ mol HBz}$$

$$\text{Molecular weight} = \frac{(\text{g HBz})}{(\text{mol HBz})} = \frac{(0.5823 \text{ g})}{(4.77 \times 10^{-3} \text{ mol})} = 122.1 \text{ g/mol},$$

so the molecular weight is 122.1 amu.

144 CHAPTER 9

9.58 See the answer in Appendix B of the text for the explanation.

9.59 See the answer in Appendix B of the text for the explanations.

9.60 See the answer in Appendix B of the text for the explanation.

9.62 The pH of the resulting solutions will depend upon the strengths of the conjugate bases that are the anions of each salt. For a polyprotic acid, the strength of the conjugate base increases with each dissociation of a proton. Thus, in this case, the sequence of dissociation reactions gives rise to the following conjugate bases in order of increasing base strength: $H_2PO_4^-$, HPO_4^{2-}, and PO_4^{3-}. Thus, the salt with the strongest conjugate base would give a solution of highest pH, etc. Highest pH is the Na_3PO_4 solution, the next highest is the Na_2HPO_4 solution, and the lowest is the NaH_2PO_4 solution.

9.64 See the answer in Appendix B of the text for the explanation.

9.66 The Henderson-Hasselbalch equation will be used.

$$pH = pK_a + \log\frac{[A^-]}{[HA]} = 3.74 + \log\frac{(1)}{(1)} = 3.74 + \log 1$$

$$= 3.74 + 0 = 3.74.$$

9.68 See the answer in Appendix B of the text for the explanations.

9.69 The Henderson-Hasselbalch equation will be used in each case.

a) $pH = pK_a + \log\frac{[A^-]}{[HA]} = 4.74 + \log\frac{(0.20\ M)}{(0.50\ M)} = 4.74 + \log(0.40)$

$$= 4.74 + (-0.40) = 4.34$$

c) $pH = pK_a + \log\frac{[A^-]}{[HA]} = 7.00 + \log\frac{(0.25\ M)}{(1.10\ M)} = 7.00 + \log(0.23)$

$$= 7.00 + (-0.64) = 6.36$$

e) $pH = pK_a + \log\frac{[A^-]}{[HA]} = 3.33 + \log\frac{(0.065\ M)}{(0.029\ M)} = 3.33 + \log(2.24)$

$= 3.33 + 0.35 = 3.68$

9.71   See answer in Appendix B of the text for the explanation.

SELF-TEST QUESTIONS

Multiple Choice

1. The Arrhenius definition of a base focuses on
   a) the acceptance of H^+
   b) the formation of covalent bonds
   c) the production of OH^-
   d) more than one of the above responses is correct

2. The pH of a 0.100 M solution of the weak acid HA is 4.91. Make appropriate assumptions, and evaluate K_a.
   a) 1.51×10^{-9}
   b) 1.23×10^{-4}
   c) 1.51×10^{-10}
   d) 6.61×10^{8}

3. A beaker contains 100 mL of a liquid with a pH of 7.0. When 0.5 mL of 0.2 M acid is added, the pH changes to 6.88. When 0.5 mL of 0.1 M base is added to another 100 mL sample of the liquid, the pH changes to 7.20. The liquid in the beaker is
   a) water
   b) an acid solution
   c) a base solution
   d) a buffer solution

The following reaction refers to Questions 4 through 7: $HCl + NaOH \rightarrow NaCl + H_2O$

4. The above reaction between hydrochloric acid and sodium hydroxide is correctly classified as
   a) combustion
   b) dehydration
   c) neutralization
   d) more than one response is correct

146   CHAPTER 9

5. The HCl solution is prepared to be 0.120 M. A 20.00 mL sample requires 18.50 mL of NaOH solution for complete reaction. What is the molarity of the NaOH solution?
   a) 0.130   b) 0.110   c) 0.0800   d) 0.200

6. How many moles of HCl would be contained in the 20.00 mL sample used in Question 5?
   a) .204   b) 2.40   c) 0.120   d) .0024

7. How many grams of HCl would be contained in the 20.00 mL sample used in Question 5?
   a) .0875   b) 4.38   c) 8.75   d) 36.5

8. A 25.00 mL sample of monoprotic acid is titrated with a standard 0.100 M base. Exactly 20.00 mL of base is required to titrate to the proper end point. What is the molarity of the acid?
   a) 0.0500   b) 0.100   c) 0.0800   d) 0.125

9. A certain solution has a pH of 1. This solution is best described as
   a) very basic   c) slightly acidic
   b) neutral   d) very acidic

10. If the pH of an aqueous solution cannot be changed significantly by adding small amounts of strong acid or strong base, the solution contains
    a) an indicator   c) a protective colloid
    b) a buffer   d) a strong acid and a strong base

True-False

11. The terms *weak acid* and *dilute acid* can be used interchangeably.

12. H_3O^+ ions are present in pure water.

13. One of the products formed in the titration of an acid by a base is water.

14. A solution with a pH of 3 is correctly classified as acidic.

15. A solution with a pH of 6 is 6 M in H^+ ions.

16. Three different sodium salts of phosphoric acid, H_3PO_4, are possible.

17. The anion produced by the first step in the dissociation of sulfurous acid, H_2SO_3, is SO_3^{2-}.

18. The second step in the dissociation of H_3PO_4 produces $H_2PO_4^-$.

19. In pure water, $[H_3O^+] = [OH^-]$.

20. In an acid-base titration, the point at which the acid and base have exactly reacted is called the equivalence point.

21. The pH is 7 at the equivalence point of all acid-base titrations.

Matching

Match the classifications given on the right to the species listed on the left. The species are involved in the following reversible reaction. Responses can be used more than once.

$$Na^+ + C_2H_3O_2^- + H_2O \rightleftarrows Na^+ + HC_2H_3O_2 + OH^-$$

22. Na^+
23. $C_2H_3O_2^-$
24. H_2O
25. $HC_2H_3O_2$

a) behaves as a Brønsted acid
b) behaves as a Brønsted base
c) behaves as neither a Brønsted acid nor base
d) behaves as both a Brønsted acid and base

Choose the response that best completes each reaction below. The acid involved in each reaction is represented by HA.

26. $2HA +$ _____ $\rightarrow H_2 + MgA_2$

27. $HA + H_2O \rightarrow$ _____ $+ A^-$

28. $HA + NaOH \rightarrow H_2O +$ _____

29. $2HA +$ _____ $\rightarrow CO_2 + H_2O + 2NaA$

a) NaA
b) Na_2CO_3
c) Mg
d) H_3O^+

Choose a silver compound formula from the right to complete each of the reactions used to prepare silver salts. In each reaction, HA represents an acid.

30. 2HA + _____ → 2AgA + H$_2$O

31. 2HA + _____ → 2AgA + CO$_2$ + H$_2$O

32. HA + _____ → AgA + H$_2$O

33. HA + _____ → AgA + CO$_2$ + H$_2$O

a) AgOH
b) Ag$_2$O
c) AgHCO$_3$
d) Ag$_2$CO$_3$

Classify the systems described on the left into one of the pH ranges given as responses.

34. the [OH$^-$] = [H$_3$O$^+$] in pure H$_2$O

35. oven cleaners are strongly basic

36. the active ingredient in the stomach, digestive juice, is 0.1 M hydrochloric acid

37. a carbonated soft drink has a tart taste

a) the pH is much lower than 7
b) the pH is much higher than 7
c) the pH is near 7
d) the pH is exactly 7

Choose a description from the right that best characterizes the solution made by dissolving in water each of the salts indicated on the left.

38. Na$_2$SO$_4$

39. Na$_3$PO$_4$

40. NH$_4$Cl

a) it is acidic because hydrolysis has occurred
b) it is basic because hydrolysis has occurred
c) it is neutral because no hydrolysis occurred
d) more than one of the above is correct

ANSWERS TO PROGRAMMED REVIEW

9.1 a) dissociates   b) hydrogen (or H$^+$)   c) dissociates
     d) hydroxide (or OH$^-$)

Acids, Bases and Salts   149

9.2 a) hydrogen b) proton (H$^+$) c) proton (H$^+$) d) conjugate

9.3 a) auto b) acids c) bases d) hydronium (or H$_3$O$^+$)
     e) hydroxide (or OH$^-$) f) hydronium (or H$_3$O$^+$) g) hydroxide (or OH$^-$)

9.4 a) sour b) hydronium (or H$_3$O$^+$) c) oxides d) hydroxides
     e) carbonates f) bicarbonates g) hydrogen h) activity series

9.5 a) acid b) base c) water d) salt e) neutralization
     f) fats g) oils

9.6 a) crystalline b) cation c) anion d) hydrates
     e) water of hydration f) positive charges

9.7 a) negative logarithm b) molar c) hydrogen (or H$^+$)
     d) basic (or alkaline) e) acidic

9.8 a) dissociate b) completely c) hydrogens (or H$^+$) d) weak

9.9 a) Titration b) base (or basic) c) equivalence d) indicator
     e) equivalence

9.10 a) seven

9.11 a) water b) pH c) water d) basic (or alkaline)

9.12 a) pH b) acids c) bases d) pH e) buffer capacity

ANSWERS TO SELF-TEST QUESTIONS

1.	c	15.	F	29.	b
2.	a	16.	T	30.	b
3.	d	17.	F	31.	d
4.	c	18.	F	32.	a
5.	a	19.	T	33.	c
6.	d	20.	T	34.	d
7.	a	21.	F	35.	b
8.	c	22.	c	36.	a
9.	d	23.	b	37.	a
10.	b	24.	a	38.	c
11.	F	25.	a	39.	b
12.	T	26.	c	40.	a
13.	T	27.	d		
14.	T	28.	a		

CHAPTER 10

Organic Compounds: Alkanes

PROGRAMMED REVIEW

Section 10.1 Carbon: The Element of Organic Compounds

Wöhler's synthesis of the compound (a) _____ led to the downfall of the vital force theory. The element (b) _____ is present in all organic compounds. Elements and compounds not studied in organic chemistry are considered to be a part of (c) _____.

Section 10.2 Organic and Inorganic Compounds Compared

The number of known organic compounds is more than (a) _____. The bonding forces normally present within organic molecules are (b) _____. The solubility of organic compounds in water is often (c) _____. (d) _____ compounds are usually nonflammable.

Section 10.3 Bonding Characteristics and Isomerism

Four (a) _____ hybrid orbitals of carbon form bonds with hydrogen atoms in CH_4. Compounds with the same molecular formula but with the atoms bonded in different patterns are called (b) _____. The number of covalent bonds normally surrounding a carbon atom is (c) _____. The number of covalent bonds normally surrounding an oxygen atom is (d) _____.

Section 10.4 Functional Groups: The Organization of Organic Chemistry

Organic compounds are arranged into classes on the basis of structural features called (a) _____ _____. Structural formulas that show all covalent bonds are referred to as (b) _____. Structural formulas that show only certain covalent bonds are referred to as (c) _____.

Section 10.5 Alkane Structures

Organic compounds containing only the elements carbon and hydrogen are called (a) _____. Saturated hydrocarbons are also known as (b) _____. An alkane with six carbon atoms has (c) _____ hydrogen atoms. Alkanes are classified as normal or branched. The compound $CH_3CH_2CH_2CH_2CH_3$ is a (d) _____ alkane.

Section 10.6 Conformations of Alkanes

Different orientations produced in a molecule by the rotation about single bonds are called (a) _____.

Section 10.7 Alkane Nomenclature

The IUPAC ending for the name of an alkane compound is (a) _____. The root word hept- is used in naming an alkane with (b) _____ carbon atoms in the longest chain. The alkyl group CH_3CH- with CH_3 substituent is called (c) _____. The alkyl group CH_3CH_2CH- with CH_3 substituent is called (d) _____. The longest carbon chain in $CH_3CHCH_2CH_2CH_3$ with CH_2CH_3 substituent has (e) _____ carbon atoms. The IUPAC name for $CH_3CHCH_2CHCH_3$ with CH_3 and CH_2CH_3 substituents is (f) _____.

Section 10.8 Cycloalkanes

The number of hydrogen atoms present in a molecule of cyclopentane with a CH_3 group is (a) _____.

A cycloalkane with 14 hydrogen atoms has (b) _____ carbon atoms. The cycloalkane cyclopentane with two CH_3 groups has groups located at positions (c) _____. The IUPAC name for cyclobutane with CH_3 and $CH_2CH_2CH_3$ groups is (d) _____.

Section 10.9 The Shape of Cycloalkanes

The most stable geometric arrangement of four atoms attached to a carbon atom is (a) _____. Another name for *cis-trans* isomers is (b) _____ isomers. The prefix (c) _____ is used to denote an isomer in which two groups are attached to the same side of a cycloalkane ring. The IUPAC name for △⟨CH₃ / CH₃ is (d) _____.

Section 10.10 Physical Properties of Alkanes

A compound in an homologous series differs from the next member in the series by a (a) _____ unit. Hydrophobic molecules are (b) _____ in water. Alkanes are (c) _____ dense than water. Liquid alkanes generally contain from (d) _____ to 20 carbon atoms.

Section 10.11 Alkane Reactions

Alkanes are the (a) _____ reactive of all organic compounds. The products of complete combustion of an alkane are (b) _____ and H₂O. The products of incomplete combustion of an alkane may be (c) _____ or C and H₂O.

SOLUTIONS TO EXERCISES ANSWERED IN THE TEXT

10.2   Wöhler prepared the compound urea, a compound recognized as organic, from mineral substances.

10.4   a) inorganic   e) inorganic
       c) organic

10.5   covalent

10.7   a) organic. Most organic compounds are flammable.
       c) organic. Many organic compounds are insoluble in water.
       e) organic. A low-melting solid is characteristic of covalent materials.

10.8   b) solubility in water

10.10 four

10.12 tetrahedral

10.14 carbon, four; hydrogen, one; oxygen, two; nitrogen, three; bromine, one

10.15 a) 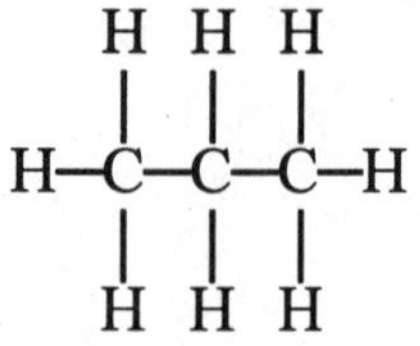
```
    H H H
    | | |
H—C—C—C—H
    | | |
    H H H
```

c)
```
  H O
  | ‖
H—C—C—H
  |
  H
```

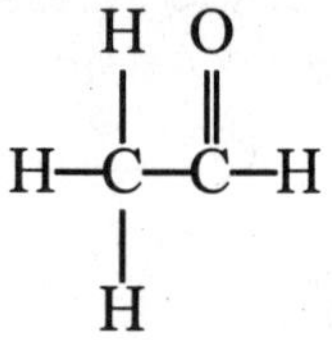

e) 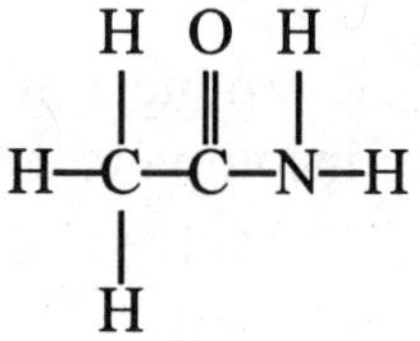
```
  H O H
  | ‖ |
H—C—C—N—H
  |
  H
```

10.16 b, c, e

10.18 a) incorrect structure because hydrogen cannot form two bonds
b) correct
c) incorrect structure because a carbon atom is shown with five bonds
d) incorrect structure because the second carbon from the left has only three bonds
e) correct

10.20 a) $CH_3CH_2CH_2-O-CH_2CH_2CH_3$

c) 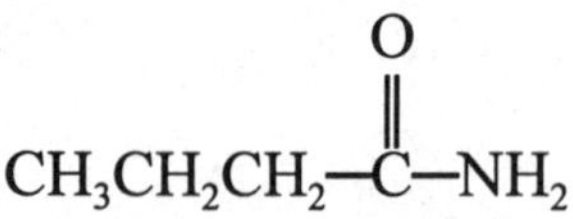 $CH_3CH_2CH_2-\overset{\overset{O}{\|}}{C}-NH_2$

10.21 a) 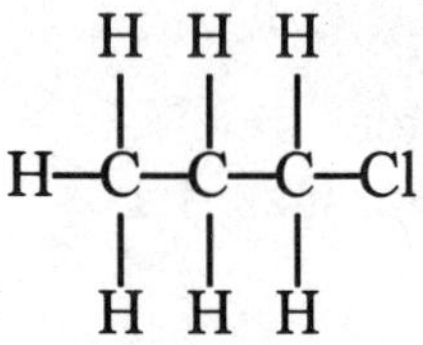
```
    H H H
    | | |
H—C—C—C—Cl
    | | |
    H H H
```

Organic Compounds: Alkanes 155

10.22 a) normal
c) branched
e) normal

10.24 b) structural
d) structural isomers

10.25 a) 7
c) 7

10.26 a) ethyl
c) n-propyl
e) isopropyl

10.27 b) 2,2-dimethylpropane
d) 3-ethylhexane
f) 5-isopropyl-2-methyloctane

10.28 a) 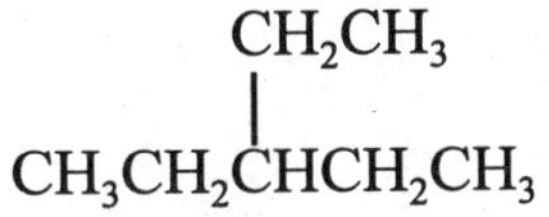 CH₃CH₂CH(CH₂CH₃)CH₂CH₃

c) 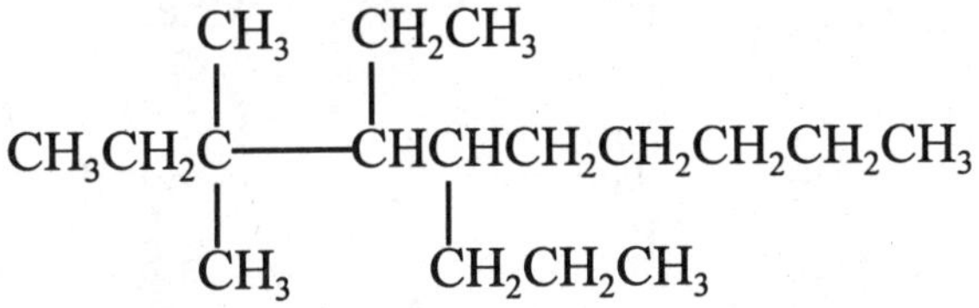 CH₃CH₂C(CH₃)(CH₃)—CH(CH₂CH₃)CH(CH₂CH₂CH₃)CH₂CH₂CH₂CH₂CH₃

e) 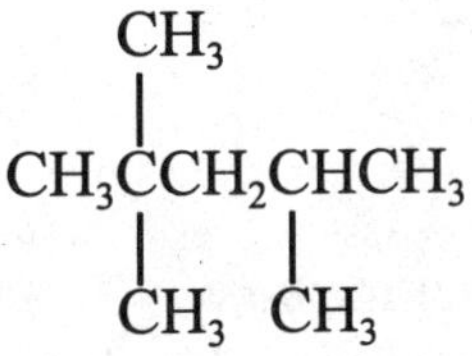 CH₃C(CH₃)(CH₃)CH₂CH(CH₃)CH₃

10.30 CH₃CH(CH₃)CH₂CH₂CH₂CH₂CH₃   2-methylheptane

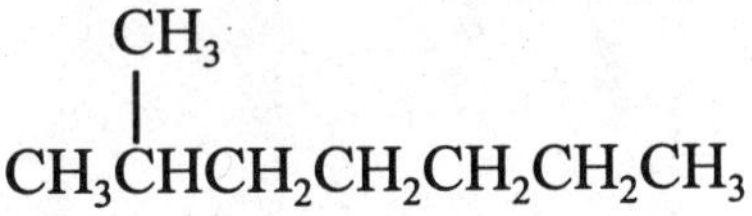

CH₃CH₂C(CH₃)(CH₃)CH₂CH₂CH₃   3,3-dimethylhexane

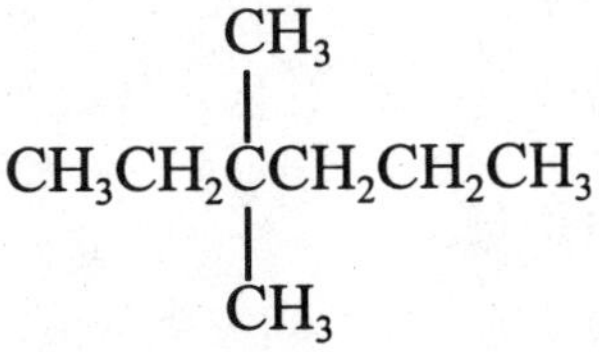

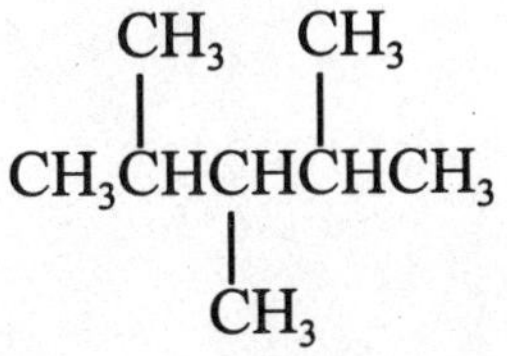

CH₃CH(CH₃)CH(CH₃)CH(CH₃)CH₃   2,3,4-trimethylpentane

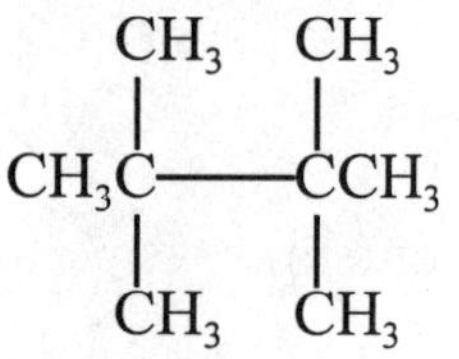

156   CHAPTER 10

$$\underset{\underset{CH_3}{|}}{\overset{\overset{CH_3}{|}}{CH_3C}}\text{———}\underset{\underset{CH_3}{|}}{\overset{\overset{CH_3}{|}}{CCH_3}}$$

2,2,3,3-tetramethylbutane

10.32 a) 4,6,6-triethyl-2,7-dimethylnonane

c) 4-*t*-butyl-5-isopropyloctane

10.33 a) 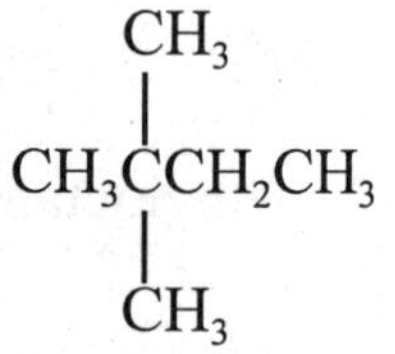 $CH_3\underset{\underset{CH_3}{|}}{\overset{\overset{CH_3}{|}}{C}}CH_2CH_3$

Both attached groups need to be located. The correct name is 2,2-dimethylbutane.

c) 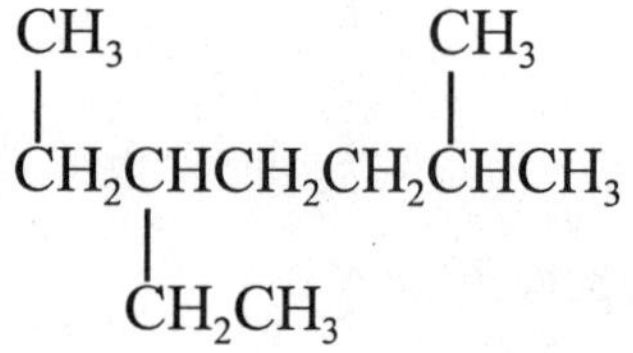 $\underset{\underset{CH_2CH_3}{|}}{CH_2\overset{\overset{CH_3}{|}}{CH}CH_2CH_2\overset{\overset{CH_3}{|}}{CH}CH_3}$

The methyl group at position 1 is a part of the longest chain, which should be heptane. The correct name is 5-ethyl-2-methylheptane.

e) 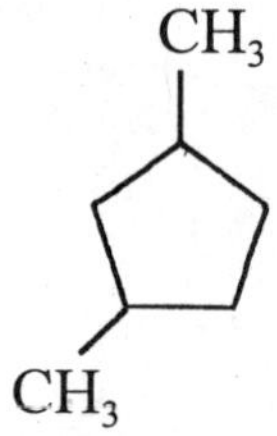 [cyclopentane with CH₃ groups at positions 1 and 3]

The ring should be numbered so that the methyl groups are at positions 1 and 3. The correct name is 1,3-dimethylcyclopentane.

10.35 a) cyclopentane
       c) 1,1-dimethylcyclohexane

e) *t*-butylcyclohexane
g) 1-ethyl-2-*n*-propylcyclopropane

10.36 b) 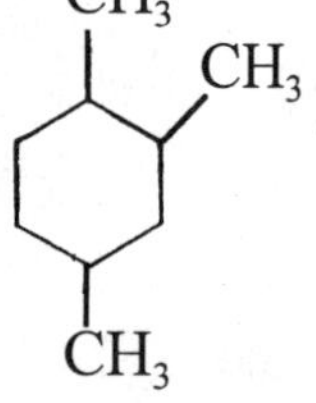 [cyclohexane with CH₃ groups]

d) 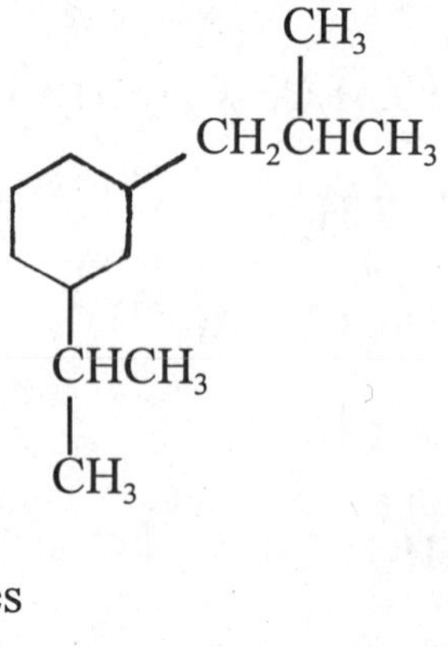 [cyclohexane with CH_2CHCH_3 (CH₃) and $CHCH_3$ (CH₃) substituents]

10.37 a) no

c) yes

Organic Compounds: Alkanes 157

10.39 [structures: cyclopentane; methylcyclobutane; 1,1-dimethylcyclopropane; 1,2-dimethylcyclopropane; ethylcyclopropane]

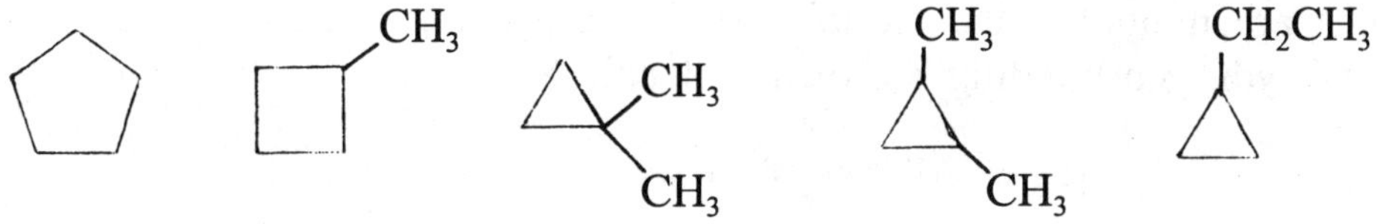

10.41 Structural isomers differ in the order of linkage of atoms. Geometric isomers differ in the three-dimensional arrangements of their atoms but not in the order of linkage of atoms.

10.42 b)

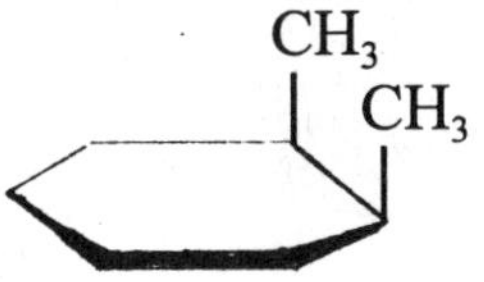

cis-1,2-dimethylcyclohexane 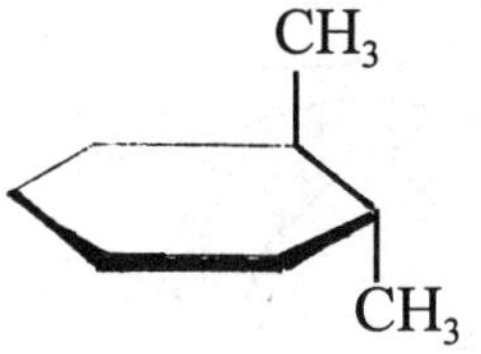   trans-1,2-dimethylcyclohexane

d)

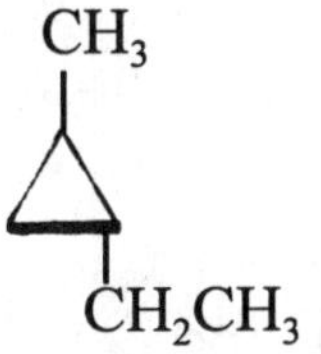

trans-1-ethyl-2-methylcyclopropane 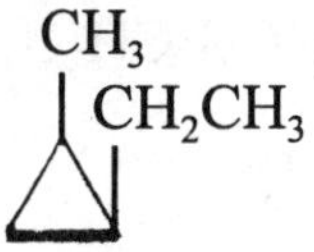   cis-1-ethyl-2-methylcyclopropane

10.44 a) *trans*-1-ethyl-2-methylcyclopropane
c) *trans*-1-methyl-3-*n*-propylcyclobutane

10.45 a) $CH_3CH_2CH_2CH_2CH_2CH_3$

c)  CH_3 CH_2 CH_3 CH_3 CH_2-CH_3
       \ / \ /   \   /
       CH_2 CH_2 CH_2-CH_2

e) $CH_3CH_2CH_2CH_2CH_2CH_3$

g)

10.47 Alkanes are nonpolar and exhibit weak interparticle forces. Water is polar and exhibits hydrogen bonding between molecules.

10.49 a)

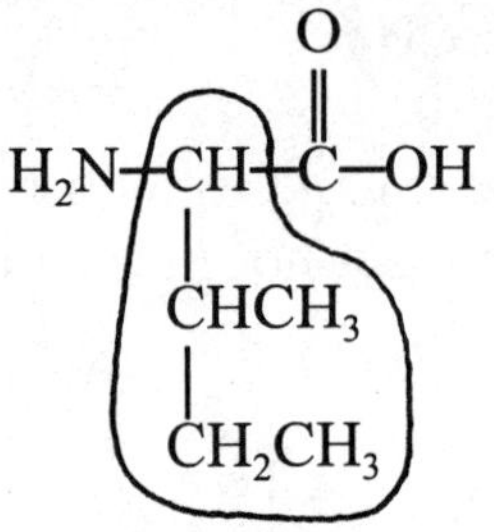

$$H_2N-\underset{\underset{\underset{CH_2CH_3}{|}}{\underset{CHCH_3}{|}}}{CH}-\underset{\underset{}{\|}}{\overset{O}{C}}-OH$$

(with H$_2$N—CH—C(=O)—OH as the backbone and the side chain CH(CH$_3$)CH$_2$CH$_3$ enclosed together with the α-carbon)

c)

$$H_2C\underset{\underset{S-S}{\diagdown\diagup}}{\overset{\diagup\hspace{-2pt}CH_2\hspace{-2pt}\diagdown}{}}CH(CH_2)_4-\overset{O}{\underset{\|}{C}}-OH$$

(ring: H$_2$C—CH$_2$—CH—(CH$_2$)$_4$— with S—S bridge; the cyclic disulfide portion is circled)

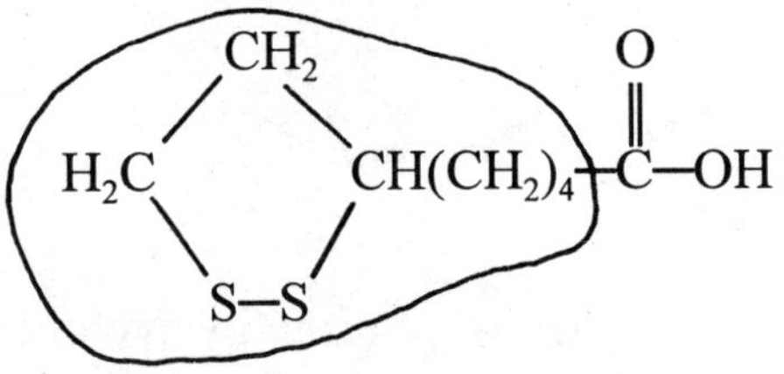

10.51 b) $CH_3CHCH_2CH_3 + 8\,O_2 \longrightarrow 5\,CO_2 + 6\,H_2O$
                |
               CH_3

10.53 The foul odor indicates escaping gas, which may explode.

10.55 Many of the essential constituents of humans and all other living matter are organic compounds.

10.57 Natural gas is an inexpensive and clean-burning fuel.

10.58 a) a fuel for heating and cooking
      c) fuel for stoves, diesel and jet engines

10.60 Carbon monoxide binds to hemoglobin, interfering with its ability to transport oxygen to the cells.

SELF-TEST QUESTIONS

Multiple Choice

1. The maximum number of covalent bonds which carbon can form is
   a) 1 b) 2 c) 3 d) 4

2. Which of the following is considered an organic compound?
   a) CH_4 b) NaOH c) Na_2CO_3 d) KCN

3. How many hydrogen atoms are needed to complete the following structure:

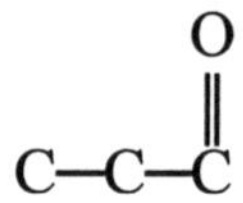

   C—C—C(=O)

   a) 2 b) 4 c) 6 d) 8

4. A C-H bond in CH_4 is formed by the overlap of what orbitals?
   a) sp^3 and 1s
   b) 1s and 1s
   c) p and 1s
   d) sp and 1s

5. A structural isomer of $CH_3CH_2CH_2$—OH is

   a) 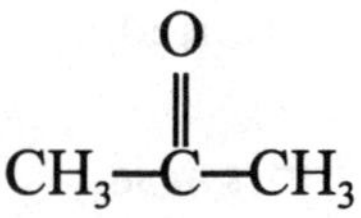 CH_3—C(=O)—CH_3
   b) CH_3CH(OH)CH_3
   c) CH_3CH_2—CH(=O)
   d) 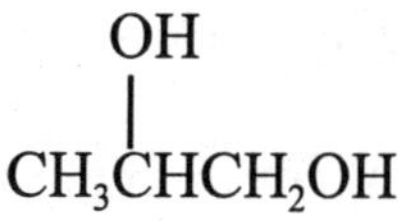 CH_3CH(OH)CH_2OH

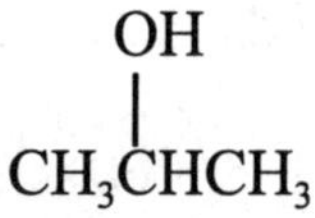

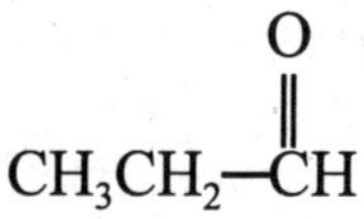

Matching

Match a class from the right to each of the structural features on the left.

6. Contains two carbon-oxygen single bonds.

7. Contains a carbon-carbon double bond.

8. Contains both an oxygen and a nitrogen.

a) alcohol
b) ether
c) alkene
d) amide

Match a class from the right to each of the compounds on the left.

9. 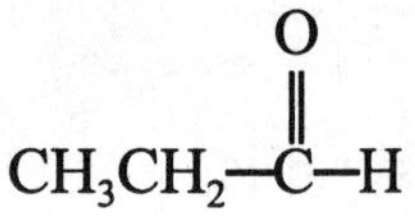 CH₃CH₂—C(=O)—H

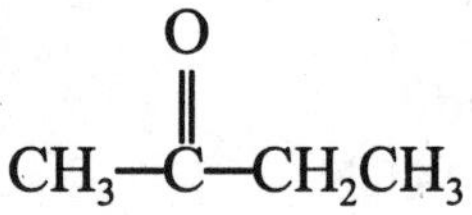

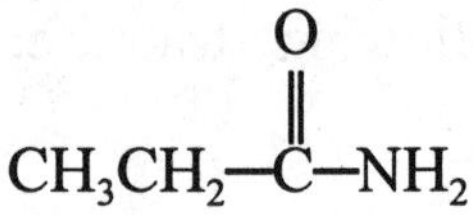

a) ether
b) aldehyde
c) amide
d) ketone

10. CH₃—C(=O)—CH₂CH₃

11. CH₃CH₂—C(=O)—NH₂

True-False

12. Most organic compounds are very soluble in water.

13. Covalent bonds are more prevalent in inorganic compounds than in organic compounds.

14. Solutions of inorganic compounds are better electrical conductors than solutions of organic compounds.

15. There are more known inorganic compounds than organic compounds.

16. A few compounds of carbon, such as CO_2, are classified as inorganic.

17. Structural isomers always have the same molecular formula.

18. Structural isomers always have the same functional group.

19. A molecule may have more than one functional group.

20. A condensed structural formula may show some bonds.

21. An expanded structural formula may not show all the bonds.

Multiple Choice

22. Which of the following compounds are identical?

    1) CH₃CH₂CH₂CH₂CH₃

    2) 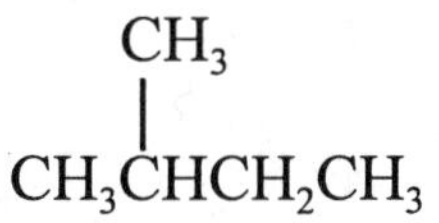 CH₃CHCH₂CH₃
         |
         CH₃

    3) 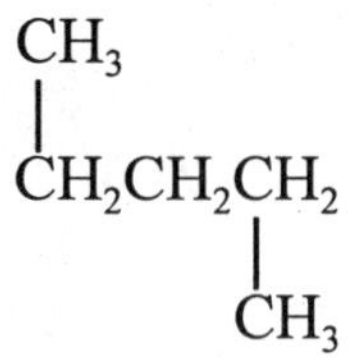 CH₂CH₂CH₂
       |
       CH₃
           |
           CH₃

    4) 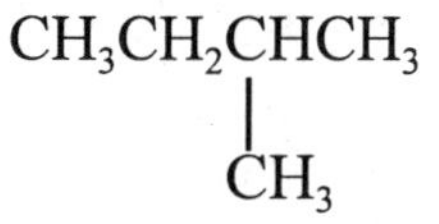 CH₃CH₂CHCH₃
                |
                CH₃

    5) 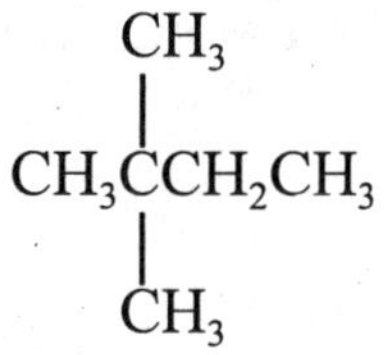 CH₃CCH₂CH₃
          |
          CH₃
       (with CH₃ above)

    a) 1 and 2   b) 1 and 3   c) 2 and 3   d) 4 and 5

23. The molecular formula for [cyclobutane with CH₃] is

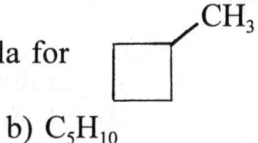

    a) C₅H₁₂   b) C₅H₁₀   c) C₅H₉   d) C₄H₉

24. Which of the following compounds is a structural isomer of CH₃CH₂CHCH₃?
                                                                      |
                                                                      CH₃

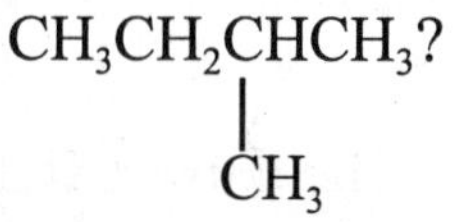

    a) CH₃CH₂CH₂CH₃

    b) 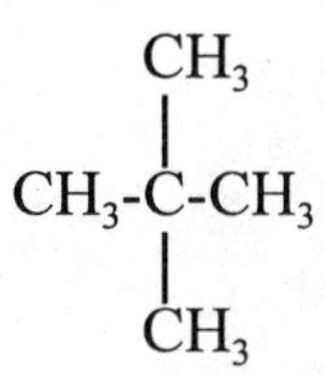 CH₃-C-CH₃
         |
         CH₃
         (with CH₃ above)

    c) CH₃CH₂CHCH₃
              |
              CH₃

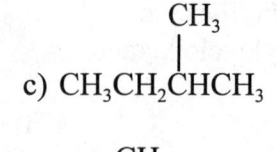

    d) 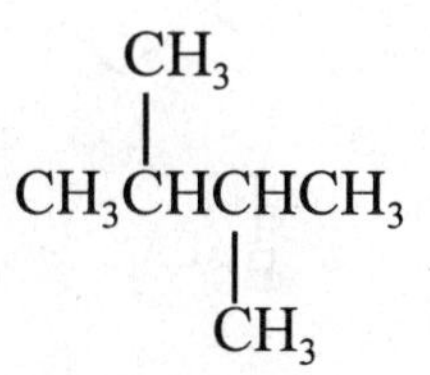 CH₃CHCHCH₃
         | |
         CH₃ CH₃

25. A structural isomer of [cyclopropane-CH₃] is

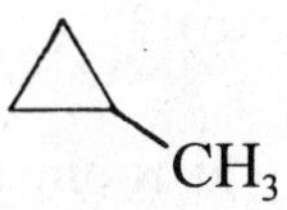

    a) [square]   b) 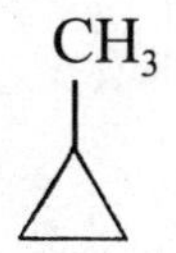 [triangle with CH₃ above]   c) CH₃CH₂CH₂CH₃   d) 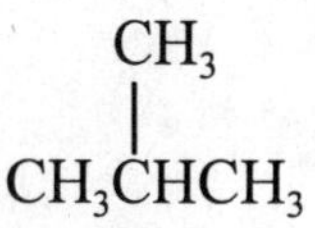 CH₃CHCH₃
                                                                            |
                                                                            CH₃

26. The normal alkane that contains eight carbon atoms is called
    a) hexane   b) heptane   c) octane   d) nonane

27. How many structural isomers have the formula C_4H_{10}?
    a) 2   b) 3   c) 4   d) 5

28. The number of carbon atoms in the longest chain of $CH_3CHCH_2\!\!\begin{array}{c}CH_3\\|\\|\\CH_2CH_3\end{array}$ is

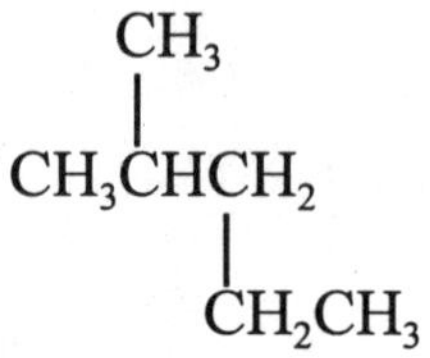

    a) 3   b) 4   c) 5   d) 6

29. The correct IUPAC name for $CH_3CHCH_2CHCH_3$ with substituents CH_2 and CH_3, where CH_2 bears a CH_3, is

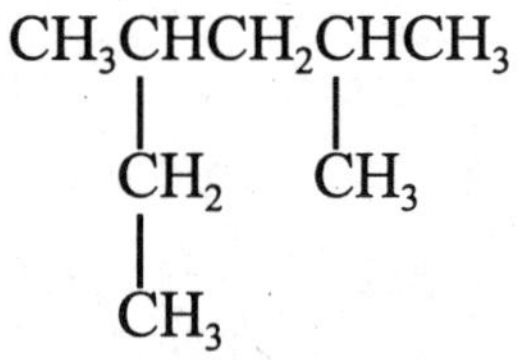

    a) 2-ethyl-4-methylpentane
    b) 2-methyl-4-ethylpentane
    c) 3,5-dimethylhexane
    d) 2,4-dimethylhexane

30. The correct IUPAC name for (methyl-substituted cyclohexane with CH_3 groups) is

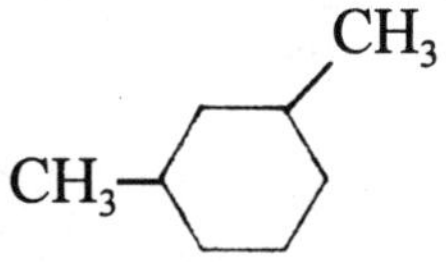

    a) 1,3-dimethylhexane
    b) 1,3-methylcyclohexane
    c) 1,5-dimethylcyclohexane
    d) 1,3-dimethylcyclohexane

31. The bromine in (methyl- and bromo-substituted cyclopentane) is located at position

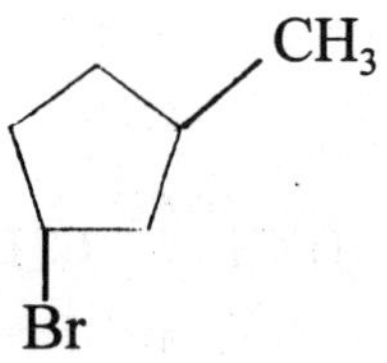

    a) 1   b) 2   c) 3   d) 5

32. A 12-carbon alkane should be a _____ at room temperature.
    a) solid   b) liquid   c) gas   d) none of these

Matching

Match an alkyl group name to each structure on the left.

33.   CH₃CH₂—

34.   CH₃CH₂CH₂—

35.   CH₃CHCH₂—
          |
          CH₃

a) *n*-propyl
b) *sec*-butyl
c) isobutyl
d) ethyl

Match the structures on the left to the descriptions on the right.

36.   [cyclopropane with CH₃ and CH₂CH₃ substituents]

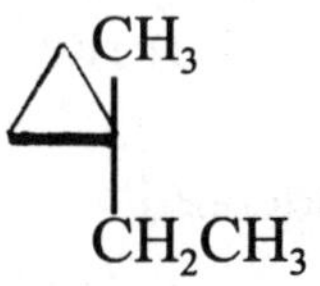

37.   [cyclobutane with CH₃ and CH₃ substituents]

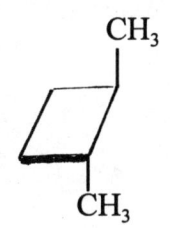

38.   [cyclopropane with CH₃ and CH₂CH₃ substituents]

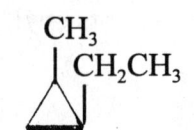

39.   [cyclohexane with CH₃ and CH₃ substituents]

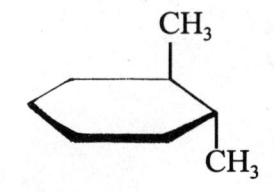

a) a *cis* compound
b) a *trans* compound
c) neither *cis* nor *trans*

True-False

40.   Pentane and cyclopentane are isomers of each other.

41.   Alkanes have lower boiling points than other organic compounds.

42.   The main component of natural gas is butane.

43. Complete combustion of pentane produces H_2O and CO_2.

44. Alkanes are polar molecules.

ANSWERS TO PROGRAMMED REVIEW

10.1 a) urea   b) carbon   c) inorganic

10.2 a) 6,000,000   b) covalent   c) low   d) inorganic

10.3 a) sp^3 b) structural isomers   c) four   d) two

10.4 a) functional groups   b) expanded   c) condensed

10.5 a) hydrocarbons   b) alkanes   c) 14   d) normal

10.6 a) conformations

10.7 a) -ane   b) 7   c) isopropyl   d) *sec*-butyl   e) 6
      f) 2,4-dimethylhexane

10.8 a) 12   b) 7   c) 1 and 3   d) 1-methyl-2-*n*-propylcyclobutane

10.9 a) tetrahedral   b) geometric   c) *cis*
      d) *trans*-1,2-dimethylcyclopropane

10.10 a) CH_2 b) insoluble   c) less   d) 5

10.11 a) least   b) CO_2 c) CO

ANSWERS TO SELF-TEST QUESTIONS

1.	d	16.	T	31.	a
2.	a	17.	T	32.	b
3.	c	18.	F	33.	d
4.	a	19.	T	34.	a
5.	b	20.	T	35.	c
6.	b	21.	F	36.	c
7.	c	22.	b	37.	b
8.	d	23.	b	38.	a
9.	b	24.	b	39.	b
10.	d	25.	a	40.	F
11.	c	26.	c	41.	T
12.	F	27.	a	42.	F
13.	F	28.	c	43.	T
14.	T	29.	d	44.	F
15.	F	30.	d		

CHAPTER 11

Unsaturated Hydrocarbons

PROGRAMMED REVIEW

Section 11.1 Nomenclature of Alkenes

A compound containing a carbon-carbon double bond has an IUPAC name with the ending (a) _____. The compound CH$_3$CH=CHCHCH$_3$ has the IUPAC name (b) _____.
                                              |
                                             CH$_3$

The IUPAC name for a compound containing two carbon-carbon double bonds has the ending (c) _____. The IUPAC name for   CH$_3$ is (d) _____.

Section 11.2 Geometry of Alkenes

Three (a) _____ hybrid orbitals are present about each carbon atom in CH$_2$=CH$_2$. The geometry of a carbon-carbon double bond and the four attached atoms is such that all six atoms lie in the same (b) _____. Geometric isomers may be designated as *cis* or *trans* compounds. The compound   H   CH$_3$ is shown as a (c) _____ geometric
                                               C=C
                                          CH$_3$ H

isomer. To complete the structure Br   in such a way that a *cis*
                                            C=C
                                       CH$_3$ H

compound is formed, (d) _____ must be attached.

Section 11.3 Properties of Alkenes

The physical properties of the alkenes are similar to those of the (a) _____. The characteristic reactions of alkenes are referred to as (b) _____ reactions. The product of CH$_3$CH$_2$CH=CH$_2$ + Br$_2$ is (c) _____. Another name for an alkyl halide is (d) _____.

168   CHAPTER 11

An alkane can be produced from an alkene by a reaction called (e) _____. According to Markovnikov's Rule, the (f) _____ carbon atom of

$$CH_3-\underset{\underset{CH_3}{|}}{C}=CHCH_3$$
$$1\quad\quad 2\quad 3\quad\, 4$$

would gain a hydrogen atom in the reaction with HCl? The chemical reaction in which a molecule of water adds to a carbon-carbon double bond is called (g) _____.

Section 11.4 Addition Polymers

The chemical reaction in which hundreds of alkene molecules react to form a large molecule is called (a) _____. The starting materials for this same process are referred to as (b) _____. A fluorine-containing polymer is (c) _____. A polymer produced by using two different starting materials is referred to as a (d) _____.

Section 11.5 Alkynes

The characteristic structural feature of alkynes is the carbon-carbon (a) _____ bond. Two (b) _____ hybrid orbitals are present about each carbon atom in H—C≡C—H. IUPAC names for alkynes end with the letters (c) _____. The simplest alkyne is well-known by the common name (d) _____. Alkynes undergo the same kinds of reactions as the (e) _____.

Section 11.6 Aromatic Compounds and the Benzene Structure

Aromatic compounds are those which contain a (a) _____ ring. Non-aromatic compounds are referred to as (b) _____ compounds. The circle within the benzene structure contains (c) _____ electrons. Each carbon atom of benzene contains three (d) _____ hybrid orbitals. The six carbons and six hydrogens of benzene are arranged in a (e) _____ geometry.

Section 11.7 Nomenclature of Benzene Derivatives

Another name for methylbenzene is (a) _____. Aminobenzene is more commonly known as (b) _____. In naming a benzene derivative with groups attached at positions 1 and 3, the prefix (c) _____ may be used. The aromatic group C_6H_5- is called a (d) _____ group.

Section 11.8 Properties and Uses of Aromatic Compounds

Aromatic compounds are nonpolar and (a) _____ in water. The typical reaction of aromatic compounds is a (b) _____ reaction. Benzene and (c) _____ are aromatic compounds which are useful as laboratory solvents. The vitamin (d) _____ is an aromatic compound. The plastic bakelite is prepared commercially from the aromatic compound (e) _____.

SOLUTIONS TO EXERCISES ANSWERED IN THE TEXT

11.1 b) alkene
      d) alkene
      f) alkene
      h) alkene

11.2 a) 2-pentene
      c) 3-methyl-1-pentene
      e) cyclopentene
      g) 4-methyl-2-pentyne
      i) 1-isopropyl-2,3-dimethylcyclopropane
      k) 3-methyl-3-n-propyl-1-hexyne
      m) 7-n-propyl-1,3,5-cycloheptatriene

11.3 a) 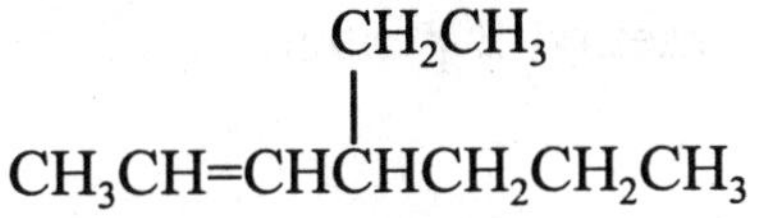 CH$_3$CH=CHCH(CH$_2$CH$_3$)CH$_2$CH$_2$CH$_3$

c) 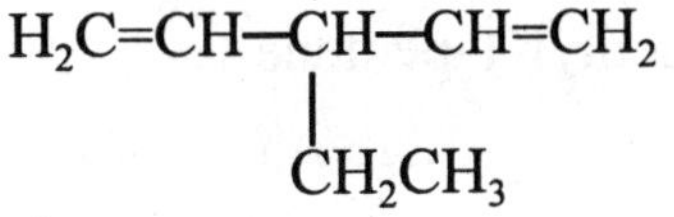 H$_2$C=CH—CH(CH$_2$CH$_3$)—CH=CH$_2$

e)  (methylcyclopropene with CH$_3$ substituent)

g)  (1,3-cyclohexadiene)

i) 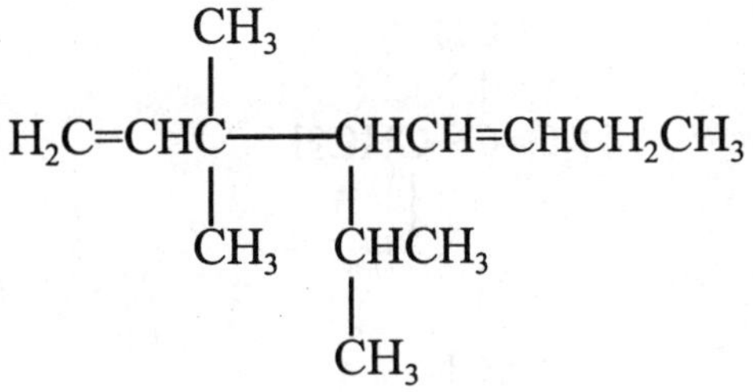 H$_2$C=CHC(CH$_3$)(CH$_3$)—CH(CH(CH$_3$)CH$_3$)CH=CHCH$_2$CH$_3$

11.5
H$_2$C=CHCH$_2$CH$_2$CH$_2$CH$_3$ 1-hexene

CH$_3$CH=CHCH$_2$CH$_2$CH$_3$ 2-hexene

CH$_3$CH$_2$CH=CHCH$_2$CH$_3$ 3-hexene

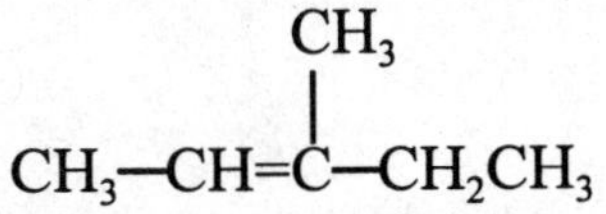

CH$_3$—CH=C(CH$_3$)—CH$_2$CH$_3$ 3-methyl-1-pentene

$$\begin{array}{c} \text{CH}_3 \\ | \\ \text{CH}_3-\text{CH}=\text{C}-\text{CH}_2\text{CH}_3 \end{array}$$ 3-methyl-2-pentene

$$\begin{array}{c} \text{CH}_3 \\ | \\ \text{H}_2\text{C}=\text{C}-\text{CH}_2\text{CH}_2\text{CH}_3 \end{array}$$ 3-methyl-1-pentene

$$\begin{array}{c} \text{CH}_3 \\ | \\ \text{CH}_3\text{C}=\text{CH}-\text{CH}_2\text{CH}_3 \end{array}$$ 2-methyl-1-pentene

$$\begin{array}{c} \text{CH}_2\text{CH}_3 \\ | \\ \text{H}_2\text{C}=\text{C}-\text{CH}_2\text{CH}_3 \end{array}$$ 2-ethyl-1-butene

$$\begin{array}{c} \text{CH}_3 \\ | \\ \text{CH}_3\text{CH}-\text{CH}=\text{CH}-\text{CH}_3 \end{array}$$ 4-methyl-2-pentene

$$\begin{array}{c} \text{CH}_3 \\ | \\ \text{CH}_3\text{CHCH}_2\text{CH}=\text{CH}_2 \end{array}$$ 4-methyl-1-pentene

$$\begin{array}{c} \text{CH}_3 \\ | \\ \text{H}_2\text{C}=\text{C}-\text{CHCH}_3 \\ | \\ \text{CH}_3 \end{array}$$ 2,3-dimethyl-1-butene

$$\begin{array}{c} \text{CH}_3 \\ | \\ \text{CH}_3-\text{C}=\text{C}-\text{CH}_3 \\ | \\ \text{CH}_3 \end{array}$$ 2,3-dimethyl-2-butene

$$\begin{array}{c} \text{CH}_3 \\ | \\ \text{H}_2\text{C}=\text{CH}-\text{CCH}_3 \\ | \\ \text{CH}_3 \end{array}$$ 3,3-dimethyl-1-butene

11.7 a) $CH_3CH_2CH=CHCH_3$ The chain should be numbered from the right, giving 2-pentene.

c) 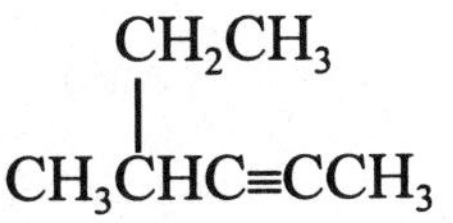 CH₃CH(CH₂CH₃)C≡CCH₃

The longest chain should be numbered from the right and should include the carbon atoms of the ethyl group. The correct name is 4-methyl-2-hexyne.

e) 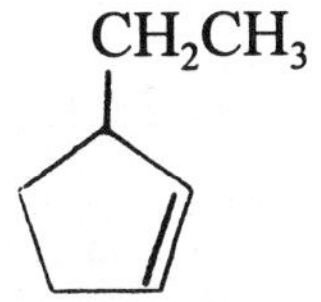 3-ethylcyclopentene structure (cyclopentene with CH₂CH₃ substituent)

The ring should be numbered so that the ethyl group is at position 3. The correct name is 3-ethylcyclopentene.

11.9 The overlap of 2p orbitals forms a π (pi) bond containing two electrons.

11.11 Structural isomers have a different order of linkage of atoms. Geometric isomers have the same order of linkage of atoms but different three-dimensional arrangements of their atoms in space.

11.13 b)

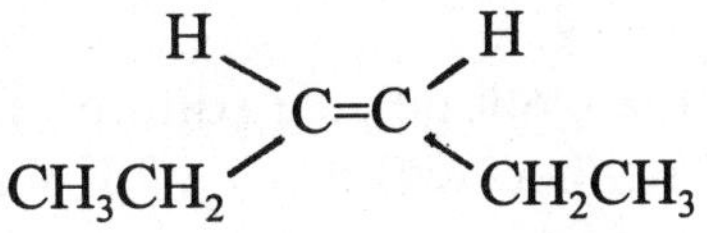

CH_3CH_2 and H on one carbon; H and CH_2CH_3 on the other — *cis*-3-hexene

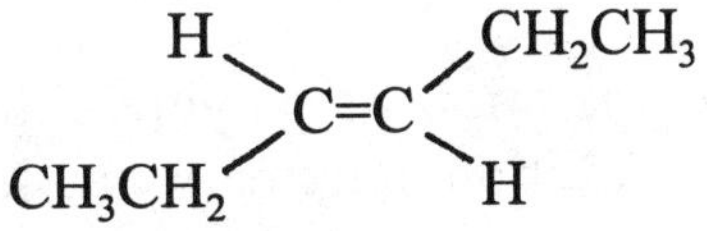

CH_3CH_2 and H on one carbon; CH_2CH_3 and H on the other — *trans*-3-hexene

d) 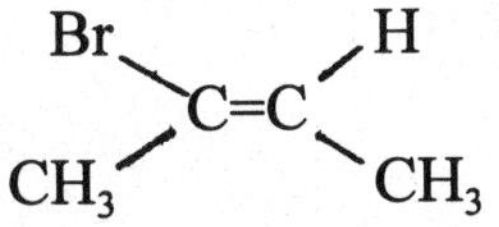 Br and CH₃ on one carbon; H and CH₃ on the other — *cis*-2-bromo-2-butene

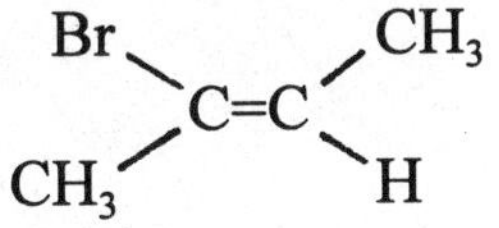

Br and CH₃ on one carbon; CH₃ and H on the other — *trans*-2-bromo-2-butene

11.14 a) 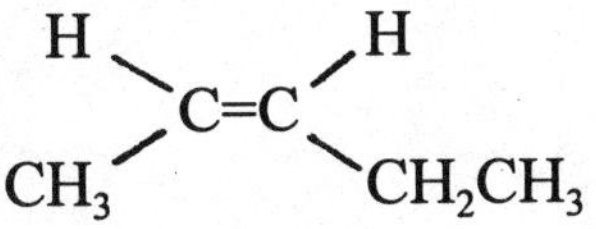 H and CH₃ on one carbon; H and CH₂CH₃ on the other

c) 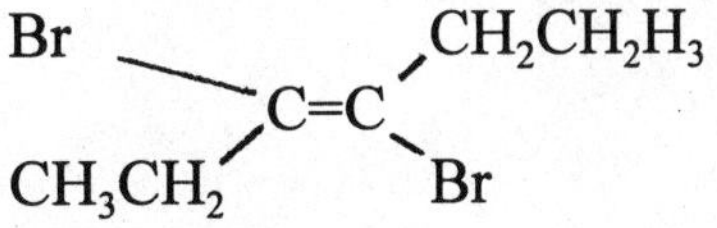 Br and CH₃CH₂ on one carbon; CH₂CH₂H₃ and Br on the other

11.16 a, c

11.18 a) [cyclopentane with CH₃ substituent]

e) [cyclohexane with OH substituent]

c) [cyclopentane]–CH(Br)–CH(Br)CH₃

g) CH₃CH(CH₃)CHCH₂CH₃ + CH₃C(CH₃)CH₂CH₂CH₃
             | |
             Br   Br

11.19 a) $CH_3CH_2CH=CHCH_3$

c) $H_3C=CHCH_3$ with CH₃ substituent or $H_2C=C(CH_3)-CH_2CH_3$

11.20 b) H_2 with Pt catalyst

d) H_2O with H_2SO_4 catalyst

11.22 The addition of Br_2 to these samples will cause the cyclohexane solution to become light orange with unreacted Br_2, whereas the 2-hexene will react with the Br_2 and remain colorless.

11.24 A monomer is a starting material used in the preparation of polymers, long-chain molecules made up of many repeating units. Addition polymers are long-chain molecules prepared from alkene monomers through numerous addition reactions. A copolymer is prepared from two monomer starting materials.

11.25 a) $-CF_2CF_2-CF_2CF_2-CF_2CF_2-$

11.27 $nCH_2=C(CH_3)CH_3 \xrightarrow{catalyst} -(-CH_2-C(CH_3)(CH_3)-)_n-$

11.29 a) insulation   c) airplane windows

11.31 one sigma and two pi bonds

11.33 The two atoms attached to a carbon-carbon triple bond lie in a straight line.

11.35 $H-C\equiv C-CH_2CH_2CH_3$ 1-pentyne

CH₃C≡C—CH₂CH₃   2-pentyne

$$\underset{\underset{CH_3}{|}}{CH_3-CH}-C\equiv C-H$$ 3-methyl-1-butyne

11.37 sp^2 orbitals; three

11.39 The circle represents the six evenly distributed electrons in the pi lobes.

11.41 The structure of limonene does not contain a benzene ring.

11.43 a) ethylbenzene   c) 1-bromo-3-ethylbenzene

11.44 b) 3-phenyl-1-pentene   d) 2,5-diphenylhexane

11.45 a) *o*-ethyltoluene   c) *m*-nitroaniline

11.46 a) 1,3-dichloro-5-nitrobenzene   d) 3-bromo-5-chlorobenzoic acid

11.47 a) [benzene ring with NH₂, CH₂CH₃ (ortho), and CH₂CH₃ (para to NH₂)]

e) [benzene ring with CH₃CH₂C(CH₃)(CH₂CH₃)— group attached]

c) [benzene ring with CH₃ and CH₂CH₃ in para positions]

11.49 nonpolar and insoluble in water

11.51 Cyclohexene readily undergoes addition reactions. Benzene resists and favors substitution reactions.

11.53 a) toluene   c) phenylalanine

11.55 ethylene

174 CHAPTER 11

11.57 polyethylene plastic, antifreeze, vinegar

11.59 A carbocation is a positively charged carbon atom with only three attached groups. It is formed when H⁺ attaches to a carbon atom of a carbon-carbon double bond.

11.61 A carcinogen is a cancer-causing chemical; polycyclic aromatic compounds.

11.63 When light strikes the retina, *cis*-retinal is converted to *trans*-retinal, triggering a chain of events that results in vision.

SELF-TEST QUESTIONS

Multiple Choice

1. A correct IUPAC name for CH₃CH=CHCH₂CH₂CH₂ is
                                              |
                                              Cl

    a) 1-chloro-2-hexene
    b) 1-chloro-4-hexene
    c) 6-chloro-2-hexene
    d) 1-chlorohexene

2. A correct IUPAC name for [cyclopentene with CH₃ group] is

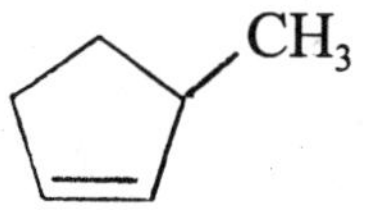

    a) 1-methylcyclopentene
    b) 3-methylcyclopentene
    c) 2-methylcyclopentene
    d) 1-methyl-2-cyclopentene

3. The correct IUPAC name for [CH₃ and H on one carbon, Cl and CH₃ on other, C=C] is

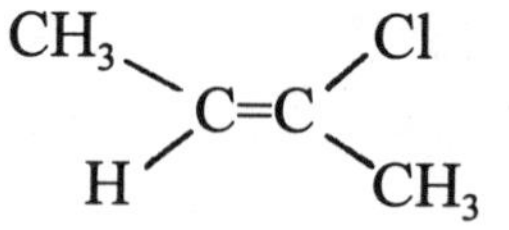

    a) 3-chloro-2-butene
    b) *trans*-3-chloro-2-butene
    c) *cis*-2-chloro-2-butene
    d) *trans*-2-chloro-2-butene

4. Which of the following can exhibit *cis-trans* isomerism?
    a) 2-methyl-1-butene
    b) 2-methyl-2-butene
    c) 2,3-dimethyl-2-butene
    d) 2,3-dichloro-2-butene

5. Markovnikov's rule is useful in predicting the product of a reaction between an alkene and
   a) H₂   b) Br₂   c) HBr   d) O₂

6. What is the product of the following reaction?

$$CH_3\underset{\underset{CH_3}{|}}{C}=CH_2 + HCl \rightarrow$$

a) $CH_3\underset{\underset{Cl}{|}}{\overset{\overset{CH_3}{|}}{C}H}CH_2\text{-}Cl$

a) CH₃CHCH₂-Cl with CH₃ on top
b) CH₃C(CH₃)(Cl)-CH₃
c) CH₃C(CH₃)-CH₂-Cl with Cl on bottom
d) CH₃CH(CH₃)CH₃

7. Which of the following is the correct representation of the polymer produced from the polymerization of H₂C=CH (with Cl substituent)?

a) ─(─H₂C─CH₂─CH(Cl)─)ₙ─

b) ─(─H₂C─C(Cl)(Cl)─)ₙ─

c) ─(─H₂C─CH(Cl)─)ₙ─

d) ─(─CH(Cl)─CH(Cl)─)ₙ─

8. A correct name for 1,2,6-trichlorobenzene is
   a) 1,2,5-trichlorobenzene
   b) 1,2,3-trichlorobenzene
   c) 1,3,5-trichlorobenzene
   d) 1,2,4-trichlorobenzene

9. A correct IUPAC name for CH₃CH(C₆H₅)─CH=CH₂ is
   a) 2-phenyl-1-butene
   b) 2-phenyl-3-butene
   c) 3-phenyl-3-butene
   d) 3-phenyl-1-butene

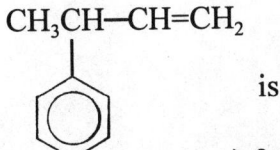

10. Addition of H₂O to CH₃CH=C—CH₃ produces
           |
          CH₃

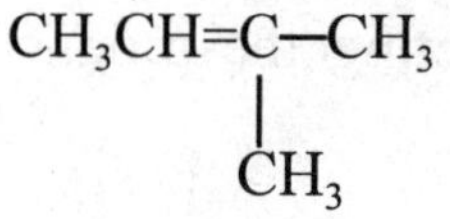

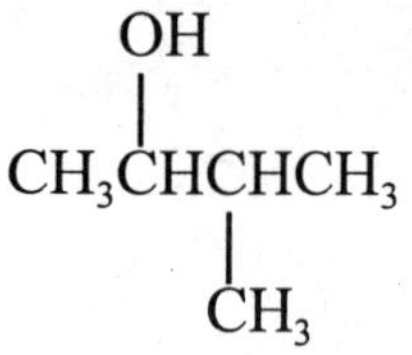

a) CH₃CHCHCH₃
      | |
     OH   CH₃

c) CH₂CH₂CHCH₃
    | |
   OH   CH₃

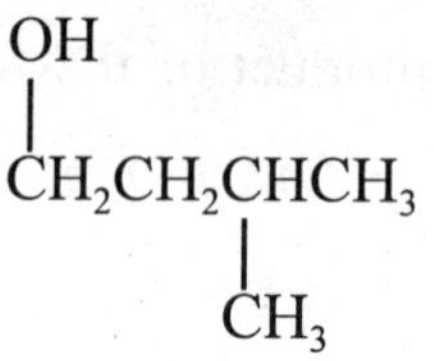

b) 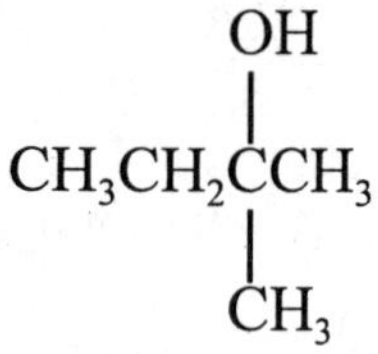 CH₃CH₂CCH₃
         | |
        OH   CH₃

d) 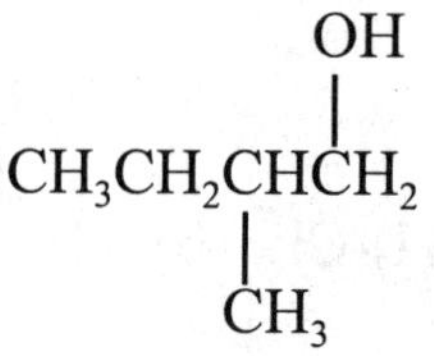 CH₃CH₂CHCH₂
         | |
        OH   CH₃

11. What reactant and catalyst are necessary to hydrogenate an alkene?
a) H⁺ and Pt
b) HCl and Pt
c) H₂ and Pt
d) H₂O and Pt

12. The IUPAC name for HC≡CCH₂CCH₃ is
                            |
                           CH₃ (above) / CH₃ (below)

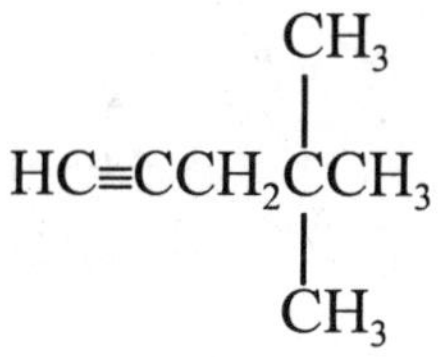

a) 2,2-dimethyl-2-pentyne
b) 2,2-dimethyl-4-pentyne
c) 4,4-dimethyl-4-pentyne
d) 4,4-dimethyl-1-pentyne

Matching

For each aromatic compound on the left, select the correct use or derived product from the responses on the right.

13. Phenol

14. Toluene

15. Aniline

16. Riboflavin

a) a solvent
b) a vitamin
c) formica
d) dyes

For each description on the left, select a correct polymer from the responses on the right.

17. Used for insulation

18. A copolymer

19. Used in airplane windows

a) saran wrap
b) plexiglass
c) polystyrene
d) PVC

True-False

20. Carbon-carbon double bonds do not rotate as freely as carbon-carbon single bonds.

21. The compound BrCH=CHCH$_3$ can exhibit *cis-trans* isomerism.

22. Alkenes are polar substances.

23. All aromatic compounds are cyclic.

24. Two hydrogens may be bonded to the same benzene carbon atom.

25. Benzene is a completely planar molecule.

26. An sp^2 hybrid orbital is obtained by mixing an *s* orbital and two *p* orbitals.

27. A copolymer is formed from two different monomers.

28. All the atoms in ethene lie in the same plane.

ANSWERS: PROGRAMMED REVIEW

11.1 a) -ene   b) 4-methyl-2-pentene   c) diene
      d) 1-methylcyclopentene

11.2 a) sp^2 b) plane   c) *trans*   d) Br

11.3 a) alkanes   b) addition   c) CH$_3$CH$_2$CHCH$_2$—Br (with Br substituent on third carbon)   d) haloalkane
      e) hydrogenation   f) third   g) hydration

178   CHAPTER 11

11.4 a) polymerization   b) monomers   c) teflon   d) copolymer

11.5 a) triple   b) *sp*   c) -yne   d) acetylene   e) alkenes

11.6 a) benzene   b) aliphatic   c) six   d) sp^2 e) planar

11.7 a) toluene   b) aniline   c) meta   d) phenyl

11.8 a) insoluble   b) substitution   c) toluene   d) riboflavin
      e) phenol

ANSWERS: SELF-TEST QUESTIONS

1.	c	11.	c	20.	T
2.	b	12.	d	21.	T
3.	d	13.	c	22.	F
4.	d	14.	a	23.	T
5.	c	15.	d	24.	F
6.	b	16.	b	25.	T
7.	c	17.	c	26.	T
8.	b	18.	a	27.	T
9.	d	19.	T	28.	T
10.	b				

CHAPTER 12

Alcohols, Phenols, and Ethers

PROGRAMMED REVIEW

Section 12.1 Nomenclature of Alcohols and Phenols

IUPAC names for alcohols end with the letters (a) _____. If two -OH groups are present in a molecule, the IUPAC name ends with the letters (b) _____. The IUPAC name for CH₃CH₂CH₂-OH is (c) _____. The structure of 2-chlorophenol is (d) _____.

Section 12.2 Classification of Alcohols

The hydroxyl-bearing carbon in a secondary alcohol is attached to (a) _____ carbon atoms. 1-Butanol is classified as a (b) _____ alcohol. Cyclopentanol is classified as a (c) _____ alcohol. If a hydroxyl-bearing carbon is attached to three carbon atoms, the compound is classified as a (d) _____ alcohol.

Section 12.3 Physical Properties of Alcohols

Alcohol molecules and water interact to form (a) _____ bonds. Alcohols with (b) _____ or fewer carbon atoms are completely soluble in water. Alcohols have higher boiling points than alkanes of similar molecular weight because of the formation of (c) _____ bonds.

Section 12.4 Reactions of Alcohols

A reaction in which water is chemically removed from an alcohol is termed (a) _____. Reaction of an alcohol with sulfuric acid at 180°C produces an (b) _____. Oxidation of a secondary alcohol produces a (c) _____. Primary alcohols undergo oxidation to (d) _____ which can be further oxidized to (e) _____.

Section 12.5 Important Alcohols

A major industrial source of formaldehyde is the alcohol (a) _____. Solutions which are referred to as tinctures contain the alcohol (b) _____. The products of fermentation of sugars are ethanol and (c) _____. The IUPAC name for rubbing alcohol is (d) _____.

179

Section 12.6 Characteristics and Uses of Phenols

A phenol used in mouthwashes and throat lozenges is (a) _____. A substance which prevents another substance from being oxidized is called an (b) _____. A popular cleaner and disinfectant for walls is (c) _____. A common additive to vegetable oils to prevent rancidity is (d) _____.

Section 12.7 Ethers

The common name for CH_3-O-$CH_2CH_2CH_3$ is (a) _____. The IUPAC name for CH_3-O-$CH_2CH_2CH_3$ is (b) _____. A ring containing an atom other than carbon is referred to as (c) _____.

Section 12.8 Properties of Ethers

Ethers are (a) _____ soluble in water than hydrocarbons of comparable molecular weight. In terms of reactivity, ethers are quite (b) _____. The boiling points of ethers are similar to those of the (c) _____.

Section 12.9 Thiols

The functional group of a thiol is (a) _____. Compounds containing the group —S—S— are referred to as (b) _____. An example of an ion which can react with thiols is (c) _____. Disulfides can be converted to thiols by the use of (d) _____.

SOLUTIONS TO EXERCISES ANSWERED IN THE TEXT

12.1 b) 1-butanol
      d) 3,4-dibromo-2-butanol
      f) 2-isopropyl-1-methylcyclopropanol
      h) 2-phenyl-1-ethanol
      j) 2-bromo-6-methylcyclohexanol
      l) 1,2-cyclopentanediol

12.2 b) isopropyl alcohol   d) ethylene glycol

12.4 b)
$$CH_3\underset{\underset{CH_3}{|}}{\overset{\overset{Br\ OH}{|\ \ |}}{C}H}CCH_2CH_3$$

d) cyclohexane with OH, CH(CH_3)$_2$ substituent (isopropyl), and Br substituent

Alcohols, Phenols and Ethers 181

12.5 b) 3,5-dinitrophenol d) 2,4,6-trimethylphenol

12.6 b)
2-(propan-2-yl)-4-(propan-2-yl)phenol structure: benzene ring with OH, CH(CH₃)₂ (as CHCH₃ with CH₃ above), and CH₃CH(CH₃) substituents.

12.8 b) primary f) tertiary
 d) secondary

12.10 Alcohols exhibit hydrogen bonding.

12.11 b) butane, 1-propanol, ethylene glycol

12.12 a) 2-Butanol because it forms hydrogen bonds with water.

12.13 a) CH₃CH₂CH₂—O···H—O—CH₂CH₂CH₃ (with H on the other side of each O)

12.14 b) CH₃CH=C(CH₃)CH₃ d) 1-methylcyclopentene

12.15 b) CH₃CH₂CH₂CH₂—O—CH₂CH₂CH₂CH₃

 d) C₆H₅—CH₂—O—CH₂—C₆H₅

12.17 b) CH₃CH₂CH(OH)CH(CH₃)CH₃

c) cyclopentyl-CH₂—OH

12.18 b) 2CH₃CH₂CH₂—OH $\xrightarrow[140°C]{H_2SO_4}$ CH₃CH₂CH₂—O—CH₂CH₂CH₃ + H₂O

d) CH₃CH₂CH(OH)CH₂CH₃ $\xrightarrow[180°C]{H_2SO_4}$ CH₃CH=CHCH₂CH₃ + H₂O

12.19 b) cyclopentene + H₂O $\xrightarrow{H_2SO_4}$ cyclopentanol (OH)

2 cyclopentanol-OH $\xrightarrow[140°C]{H_2SO_4}$ dicyclopentyl ether + H₂O

d) CH₃CH₂CH₂CH₂(OH) $\xrightarrow{H_2SO_4}$ CH₃CH₂CH=CH₂ + H₂O

CH₃CH₂CH=CH₂ $\xrightarrow{H_2SO_4}$ CH₃CH₂CH(OH)CH₃

CH₃CH₂CH(OH)CH₃ + (O) → CH₃CH₂C(=O)CH₃ + H₂O

12.21 Methanol is toxic, and even small amounts absorbed through the skin are hazardous to health.

12.23 As a moistening agent, glycerol keeps candies from drying out and hardening.

Alcohols, Phenols and Ethers

12.24 b) ethanol
      d) isopropyl alcohol
      f) ethanol

12.25 b) hexylresorcinol
      d) phenol

12.26 a) vitamin E

12.27 b) t-butyl ethyl ether
      d) diphenyl ether

12.28 b) ethoxycyclopentane
      d) 2-n-propoxyhexane

12.29 b) 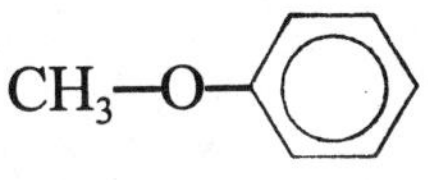 $CH_3-O-C_6H_5$
      d) 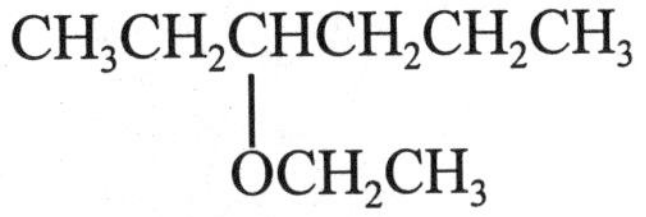 $CH_3CH_2CH(OCH_2CH_3)CH_2CH_2CH_3$

12.31 inertness

12.33 $CH_3CH_2CH_2-OH$, $CH_3CH_2-O-CH_2CH_3$, $CH_3CH_2CH_2CH_3$

This order is based on the decreasing ability to form hydrogen bonds with water.

12.35

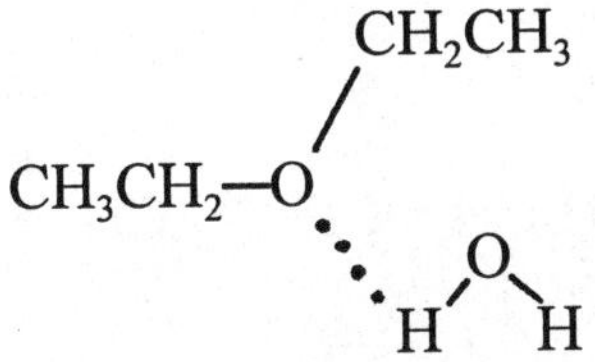

12.36 b) 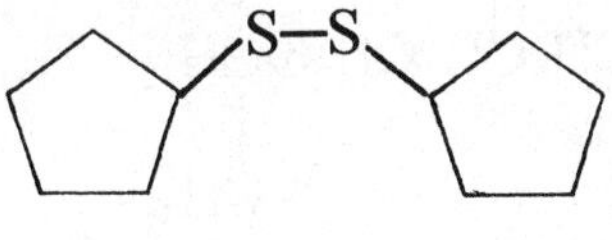 cyclopentyl-S-S-cyclopentyl + H_2O

12.38 Alcohols are oxidized to aldehydes and ketones, whereas thiols are oxidized to disulfides.

12.40

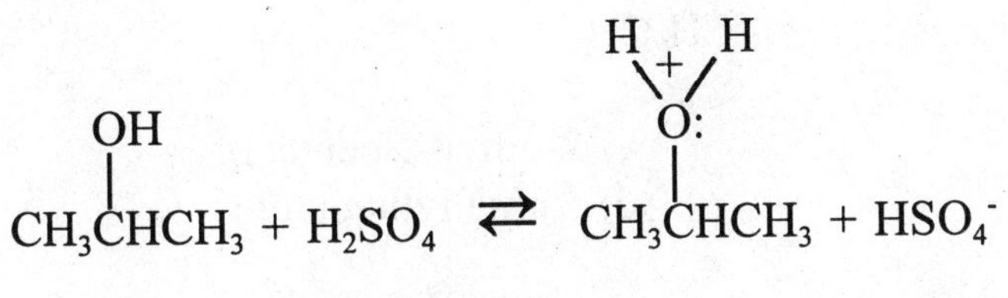

$$CH_3\overset{OH}{\underset{|}{CH}}CH_3 + H_2SO_4 \rightleftarrows CH_3\overset{\overset{+}{O}H_2}{\underset{|}{CH}}CH_3 + HSO_4^-$$

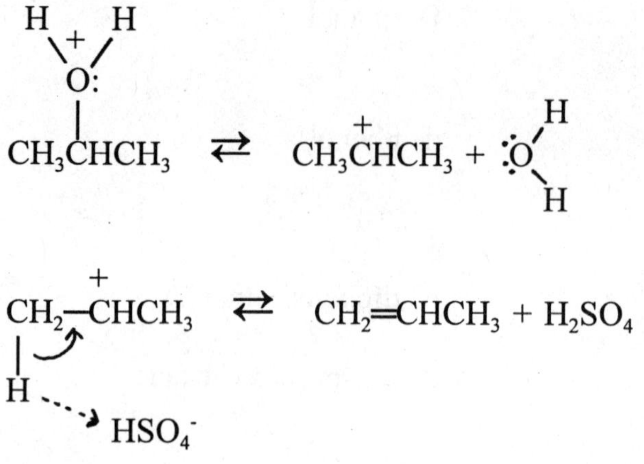

184 CHAPTER 12

$$\underset{CH_3\overset{|}{C}HCH_3}{\overset{H\overset{+}{\diagdown}/H}{O:}} \rightleftarrows \underset{CH_3\overset{+}{C}HCH_3}{} + :\underset{H}{\overset{H}{\diagup}}\overset{+}{O}$$

$$\underset{\underset{\overset{\downarrow}{HSO_4^-}}{H\cdots}}{\overset{+}{CH_2-CHCH_3}} \rightleftarrows CH_2{=}CHCH_3 + H_2SO_4$$

12.42 ethanol

SELF-TEST QUESTIONS

Multiple Choice

1. The structure of 1-*n*-propoxypropane is

   a) $CH_3CH_2OCH_2CH_2CH_3$ c) $CH_3CH_2CH_2OCH_2CH_2CH_3$

   b) $CH_3CH_2O\overset{\overset{\displaystyle CH_3}{|}}{C}H\text{-}CH_3$ d) $CH_3CH_2CH_2O\overset{\overset{\displaystyle CH_3}{|}}{\underset{\underset{\displaystyle CH_3}{|}}{C}}H$

2. A correct IUPAC name for $CH_3CH_2CH_2\overset{\overset{\displaystyle CH_3}{|}}{\underset{\underset{\displaystyle CH_2CH_3}{|}}{C}}\text{-}OH$ is

   a) 1-ethyl-1-methylbutanol   c) 2-ethyl-2-pentanol
   b) 3-methyl-3-hexanol   d) 3-methylheptanol

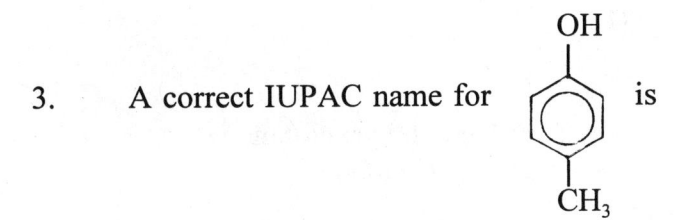

3. A correct IUPAC name for ![structure: phenol with CH3 at para position] is

   a) 4-methylphenol
   b) 1-methylphenol
   c) 2-methyl-4-phenol
   d) 4-methyl-2-phenol

4. The alcohol ![cyclopentanol structure] is

   a) primary
   b) secondary
   c) tertiary
   d) quaternary

5. Which of the following is a primary alcohol?
   a) 1-butanol
   b) 2-butanol
   c) 2-propanol
   d) 2-methyl-2-propanol

6. Which of the following is the most soluble in water?
   a) $CH_3CH_2CH_2-O-CH_3$
   b) $CH_3CH_2-O-CH_2CH_3$
   c) $CH_3CH_2CH_2CH_2-OH$
   d) $CH_3CH_2CH_2CH_2CH_2-OH$

7. What reagent is required to carry out the reaction

   $$2CH_3-SH + \underline{} \rightarrow CH_3-S-S-CH_3 + H_2O$$

   a) NaOH
   b) (O)
   c) H^+
   d) H_2SO_4

8. The chemical reactivity of ethers is closest to the
   a) alkanes
   b) alkenes
   c) aromatics
   d) alcohols

9. A common name for ethoxybenzene is
   a) ethyl benzene ether
   b) ethyl phenyl ether
   c) ethyl benzyl ether
   d) ethoxy benzyl ether

186 CHAPTER 12

10. The IUPAC name of [cyclobutane-OCH₂CH₃] is

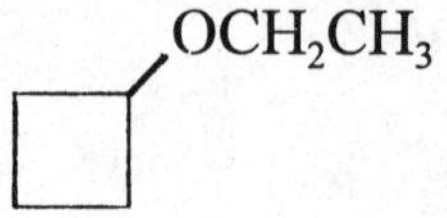

    a) ethoxycyclobutane
    b) 1-ethylcyclobutane
    c) 1-ethyloxycyclobutane
    d) ethylcyclobutoxy

Matching

For each description on the left, select the correct alcohol from the responses on the right.

11. Automobile antifreeze

12. A moistening agent in cosmetics

13. Rubbing alcohol

14. Present in alcoholic beverages

a) glycerol
b) ethylene glycol
c) isopropyl alcohol
d) ethyl alcohol

Select the correct phenol for each description on the left.

15. An antioxidant used in food packaging

16. Used in throat lozenges

17. Present in Lysol

a) *o*-phenylphenol
b) BHA
c) 4-*n*-hexylresorcinol
d) eugenol

For each reaction on the left, select the correct product from the responses on the right.

18. An alcohol is dehydrated at 140°C

19. An alcohol is dehydrated at 180°C

20. A secondary alcohol is subjected to oxidation

a) a ketone
b) an ether
c) an alkene
d) an aldehyde

True-False

21. Methyl alcohol is a product of the fermentation process.

22. Vitamin E is a natural antioxidant.

23. Tertiary alcohols undergo oxidation to produce ketones.

24. Hydrogen bonding accounts for the water solubility of certain alcohols.

25. Alcohols have a higher boiling point than ethers of similar molecular weight.

26. Oxidation of a thiol produces a disulfide.

27. Two Hg^{2+} ions are required for the reaction with a thiol.

28. Ethers can form hydrogen bonds with water molecules.

29. Thiols are responsible for the pleasant fragrance of many flowers.

30. Ethylene glycol is used as a moisturizer in foods.

ANSWERS TO PROGRAMMED REVIEW

12.1   a) ol

       b) diol

       c) 1-propanol

       d) 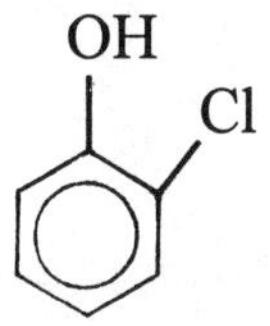 (benzene ring with OH and Cl on adjacent carbons)

12.2   a) two
       b) primary
       c) secondary
       d) tertiary

12.3   a) hydrogen
       b) three
       c) hydrogen

12.4   a) dehydration
       b) alkene
       c) ketone
       d) aldehydes
       e) carboxylic acids

12.5   a) methanol
       b) ethanol
       c) CO_2
       d) 2-propanol

12.6   a) hexylresorcinol
       b) antioxidant
       c) Lysol
       d) vitamin E

12.7 a) methyl *n*-propyl ether
　　　b) 1-methoxypropane
　　　c) heterocyclic

12.8 a) more
　　　b) unreactive
　　　c) hydrocarbons

12.9 a) SH
　　　b) disulfides
　　　c) Hg^{2+}
　　　d) reducing agent (H)

ANSWERS TO SELF-TEST QUESTIONS

1.	c	11.	b	21.	F
2.	b	12.	a	22.	T
3.	a	13.	c	23.	F
4.	b	14.	d	24.	T
5.	a	15.	b	25.	T
6.	c	16.	c	26.	T
7.	b	17.	a	27.	F
8.	a	18.	b	28.	T
9.	b	19.	c	29.	F
10.	a	20.	a	30.	F

CHAPTER 13

Aldehydes and Ketones

PROGRAMMED REVIEW

Section 13.1 Naming Aldehydes and Ketones

An aldehyde differs from a ketone in that a (a) _____ atom is attached to the carbonyl carbon. The IUPAC ending for naming an aldehyde is (b) _____. The aldehyde carbon is always located at position (c) _____. The characteristic IUPAC ending for naming a ketone is (d) _____.

Section 13.2 Physical Properties

Aldehydes and ketones have boiling points (a) _____ those of alcohols with similar molecular weights. The lack of a hydrogen atom on the oxygen prevents the formation of (b) _____ bonds in aldehydes and ketones. Aldehydes and ketones have boiling points (c) _____ those of alkanes with similar molecular weights. In terms of water solubility, low-molecular-weight aldehydes and ketones are (d) _____.

Section 13.3 Chemical Properties

Oxidation of an aldehyde produces a (a) _____. Ketones do not react when treated with (b) _____. A positive reaction with Tollens' reagent produces metallic (c) _____. Addition of hydrogen to an aldehyde or ketone requires a catalyst like (d) _____. The reaction of two molecules of alcohol with an aldehyde produces an (e) _____.

Section 13.4 Important Aldehydes and Ketones

An aldehyde used in preserving biological specimens is (a) _____. The ketone produced in the greatest quantity is (b) _____. Vanillin, the compound present in vanilla flavoring, belongs to the (c) _____ functional class.

SOLUTIONS TO EXERCISES ANSWERED IN THE TEXT

13.2 a) aldehyde
     c) ketone
     e) ketone

13.3 a) ethanol
     c) 4-methyl-2-pentanone
     e) 2-methylcyclopropanone
     g) 3-methyl-1-phenyl-2-butanone

13.4 b) 2-isobutyl-4-methoxyhexanal

13.5 a) 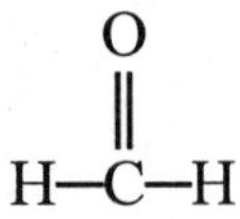 H—C(=O)—H
     c) 3-methylcyclohexanone structure
     e) H—C(=O)—CH$_2$CH(Br)CH$_2$—phenyl

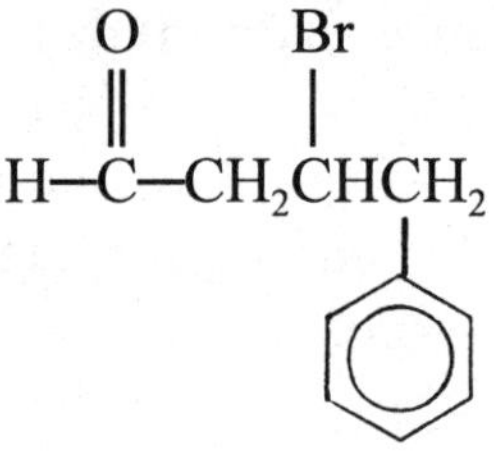

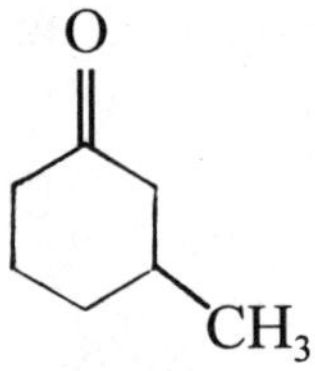

13.7 a) 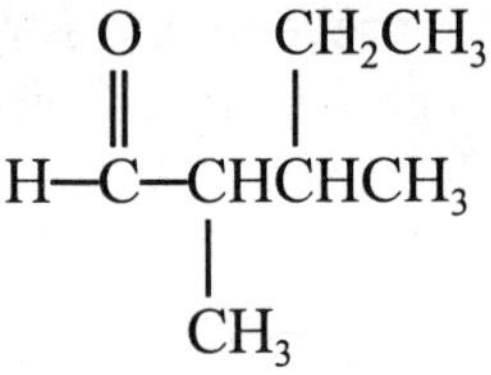 H—C(=O)—CH(CH$_2$CH$_3$)CH(CH$_3$)CH$_3$

The longest chain contains five carbon atoms. The correct name is 2,3-dimethylpentanal.

     c) 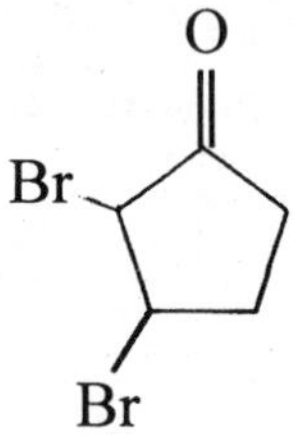 2-bromo-3-bromo cyclopentanone structure

The positions of the bromine atoms are wrong. The correct name is 2,3-dibromocyclopentanone.

13.9 Acetone dissolves remaining traces of water, and together they are discarded. The remaining small amounts of acetone evaporate.

13.11 Propane is nonpolar. Ethanol is a polar molecule and exhibits greater interparticle forces.

13.12 a) 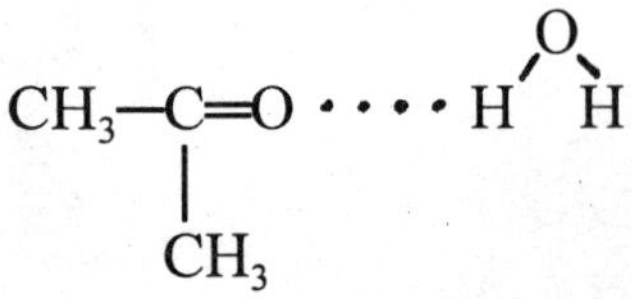 CH$_3$—C(=O)(CH$_3$) ···· H—O—H

     c) CH$_3$CH$_2$—C(=O)—H ···· H—O—H

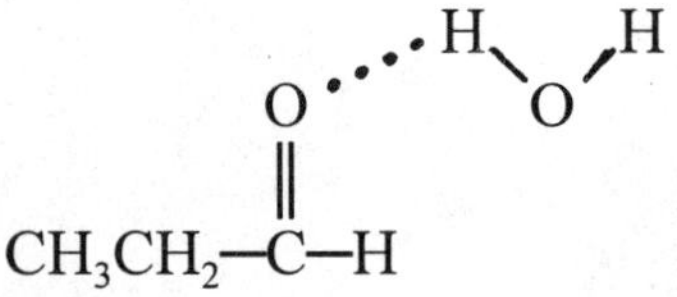

13.13 c, a, b

13.15 b)

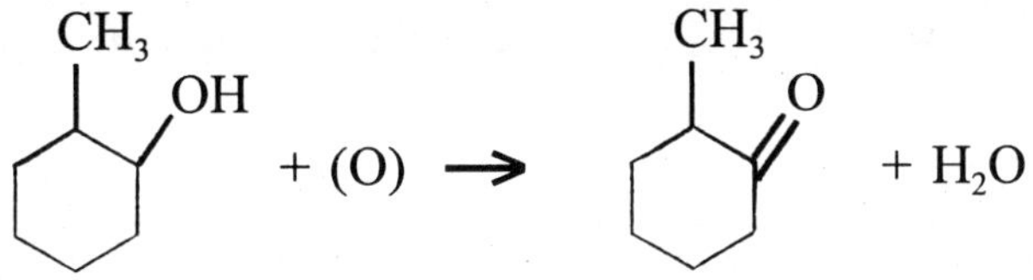

$$\underset{\text{CH}_3}{\text{cyclohexane-OH}} + (O) \longrightarrow \underset{\text{CH}_3}{\text{cyclohexanone}} + H_2O$$

13.16 a) neither   c) neither

13.17 a) acetal   e) none of these
       c) acetal

13.18 a) cyclic hemiacetal   e) hemiketal and alcohol
       c) cyclic acetal

13.19 a) aldehyde and alcohol c) hemiketal and alcohol

13.21 a) ketone   e) an aldehyde and two molecules
       c) secondary alcohol   of alcohol

13.22 the formation of a silver precipitate or mirror

13.23 a) does not react
       c) does not react

e) 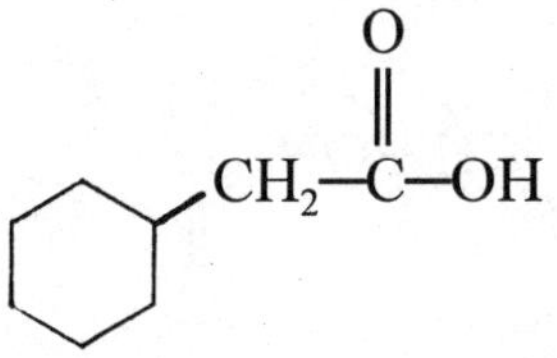 cyclohexyl–CH$_2$–C(=O)–OH

13.25 Cu$_2$O

13.26 a) no   c) yes   e) no

13.28
$$\underset{\text{OH}}{\text{CH}_2}-\underset{\text{OH}}{\text{CH}}-\underset{\text{OH}}{\text{CH}}-\underset{\text{OH}}{\text{CH}}-\underset{\text{OH}}{\text{CH}}-\underset{\text{O}}{\overset{\|}{\text{C}}}-H$$
(last two groups circled)

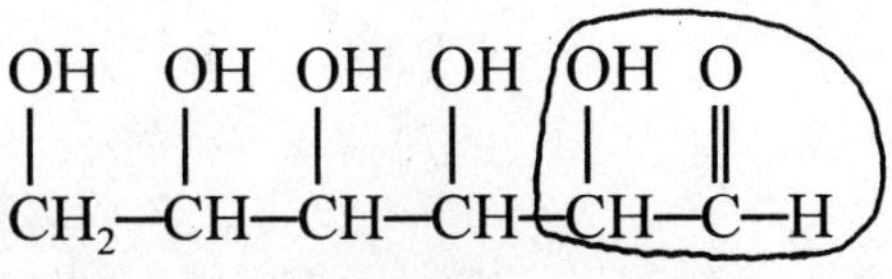

13.30
$$\underset{\text{OH}}{\text{CH}_2}-\underset{\text{OH}}{\text{CH}}-\underset{\text{OH}}{\text{CH}}-\underset{\text{OH}}{\text{CH}}-\underset{\text{O}}{\overset{\|}{\text{C}}}-\underset{\text{OH}}{\text{CH}_2}$$
or

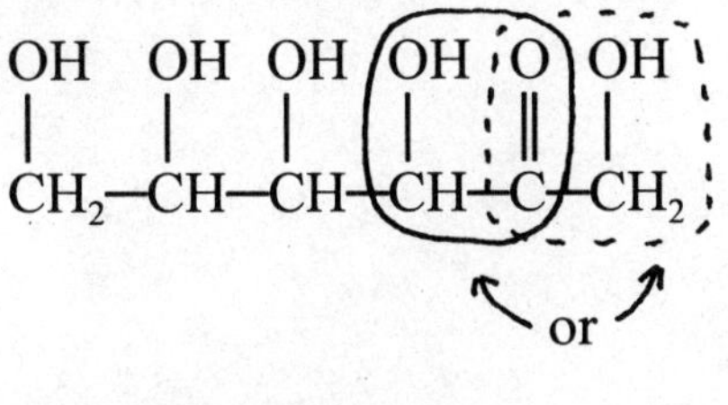

13.32 a) CH₃CH₂CH(OH)CH₃

c) [cyclohexyl]–C(OCH₃)₂–H + H₂O

e) [oxetane with OCH₂CH₃ substituent] + H₂O

g) CH₃CH₂C(OCH₂CH₂CH₃)(OCH₃)(CH₃) + H₂O

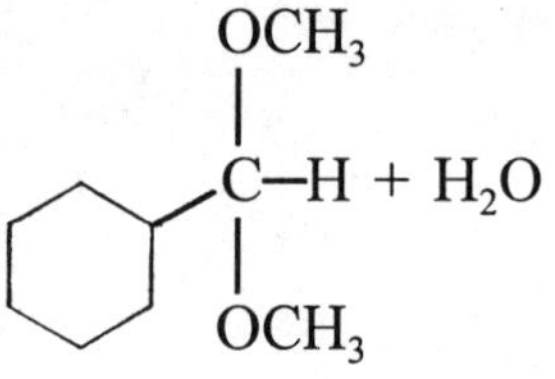

i) CH₃CH₂—OH + CH₃—C(=O)—CH₃

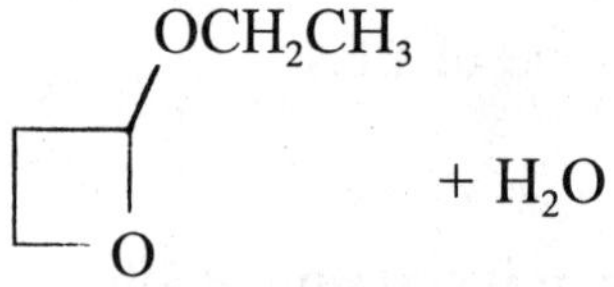

13.34 a) CH₃—C(=O)—H + CH₃CH₂—OH + CH₃—OH

c) CH₃—C(=O)—CH₃ + 2 [cyclopentanol]

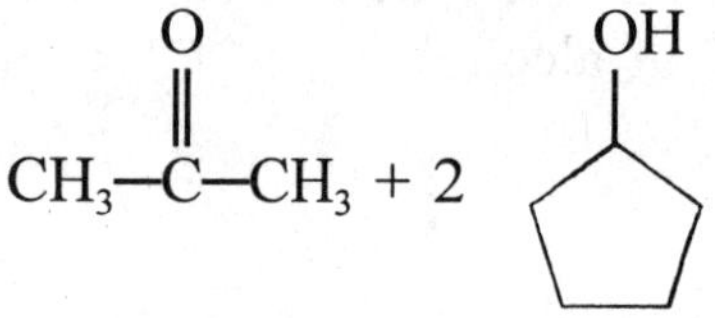

13.35 a) [cyclopentane ring with CH₂—OH and CH₂—CH₂—C(=O)H groups] + HO—[cyclopentyl]

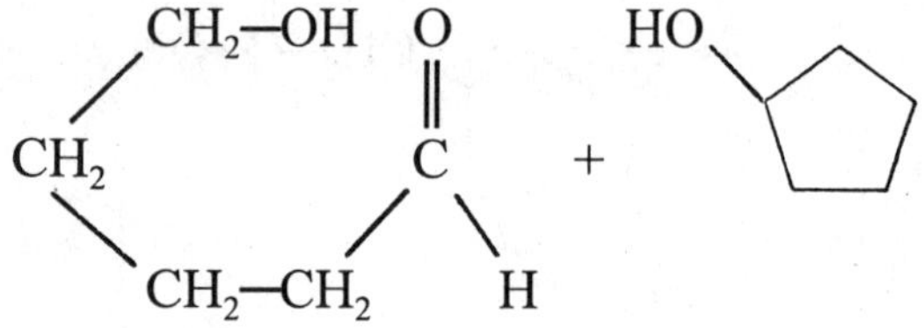

13.36 a) CH₃CH=CHCH₃ + H₂O →(H₂SO₄) CH₃CH(OH)CH₂CH₃

CH₃CH(OH)CH₂CH₃ + (O) → CH₃C(=O)CH₂CH₃ + H₂O

c) 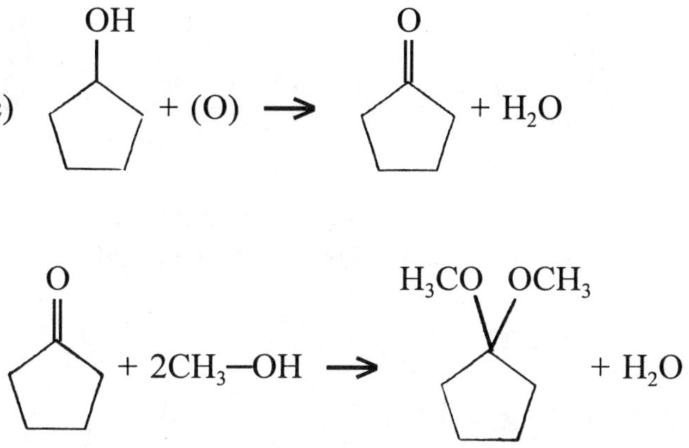 [cyclopentanol] + (O) → [cyclopentanone] + H₂O

[cyclopentanone] + 2CH₃—OH → [cyclopentane with two OCH₃ groups] + H₂O

13.38 a) menthone   e) vanillin
       c) acetone

13.39 The oxidation of ethanol produces acetaldehyde, whereas the oxidation of methanol produces formaldehyde. Formaldehyde is much more toxic than acetaldehyde.

13.41 Antabuse blocks the body's use of acetaldehyde. If a patient who has been given Antabuse consumes alcohol, acetaldehyde concentrations build up, causing nausea.

13.43 dihydroxyacetone

SELF-TEST QUESTIONS

Multiple Choice

1. A correct IUPAC name for CH₃CH₂CH₂CH₂—C(=O)—H is
    a) pentanal   c) 5-pentanal
    b) 1-pentanol   d) 1-pentanone

2. A correct IUPAC name for [cyclopentanone with ethyl group] is

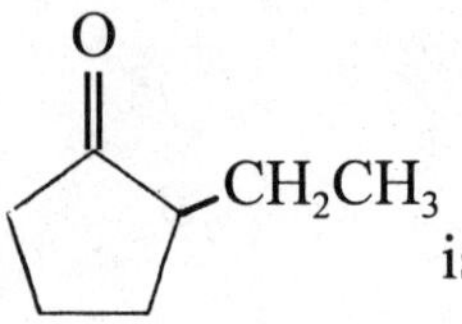

   a) 1-ethylcyclopentanone
   b) 1-ethylcyclopentanal
   c) 2-ethylcyclopentanone
   d) 2-ethylcyclopentanal

3. A correct IUPAC name for CH_2-CH_2-C-H (with Br on first carbon, =O on last) is

   a) 1-bromo-3-propanone
   b) 1-bromopropanal
   c) 3-bromopropanal
   d) 3-bromo-1-propanone

4. Which of the following pure compounds can exhibit hydrogen bonding?

   a) $CH_3CH_2-\overset{\overset{O}{\|}}{C}-H$

   b) $CH_3-\overset{\overset{O}{\|}}{C}-CH_2CH_3$

   c) $CH_3CH_2CH_2-OH$

   d) more than one response is correct

5. Which reagent may be used to test for the presence of an aldehyde?
   a) Ag^+
   b) Br_2
   c) Cu_2O
   d) NaOH

6. The reaction of water and H^+ with $CH_3\underset{\underset{OCH_3}{|}}{\overset{\overset{OCH_3}{|}}{C}}-H$ produces

   a) $CH_3\overset{\overset{O}{\|}}{C}-H + 2CH_3OH$

   b) $CH_3\underset{\underset{}{}}{\overset{\overset{OCH_3}{|}}{C}H_2} + CH_3OH$

   c) $CH_3\overset{\overset{O}{\|}}{C}-H + 2H-\overset{\overset{O}{\|}}{C}-H$

   d) $CH_3\overset{\overset{O}{\|}}{C}-H + CH_3-O-CH_3$

7. Which of the following compounds could be oxidized by Benedict's reagent?

a) CH₃CCH₂CH₃ (with C=O)

b) CH₃CH₂CH₂CH (with C=O)

c) CH₃CCHCH₃ (with C=O and OH on middle CH)
         |
         OH

d) CH₃CCH₂CH₂—OH (with C=O)

8. An important preservative for biological specimens is

a) CH₃CCH₃ (with C=O)

b) CH₃CCH₂CH₃ (with C=O)

c) HCH (with C=O)

d) CH₃CH (with C=O)

9. A flavoring for margarine is
   a) biacetyl
   b) benzaldehyde
   c) 2-butanone
   d) vanillin

10. Hydrogenation of an aldehyde produces a
    a) acetal
    b) primary alcohol
    c) secondary alcohol
    d) ketone

Matching

For each structure on the left, select the correct class of compound from the responses on the right.

11. CH₃CH₂CH(OH)—OCH₃

12. CH₃CH₂—C(OH)(CH₃)—OCH₃

13. CH₃CH₂—C(OCH₃)(CH₃)—OCH₃

14. 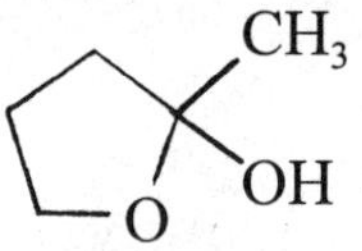 2-methyl-2-hydroxytetrahydrofuran

a) acetal
b) ketal
c) hemiacetal
d) hemiketal

Select the correct product for each of the reactions on the left.

15. aldehyde + (O) →

16. ketone + hydrogen $\xrightarrow{Pt}$

17. aldehyde + alcohol →

18. ketone + alcohol ⇌

a) hemiacetal
b) hemiketal
c) carboxylic acid
d) alcohol

True-False

19. Both aldehydes and ketones contain a carbonyl group.

20. Pure ketones can hydrogen bond.

21. Pure aldehydes can hydrogen bond.

22. Both aldehydes and ketones can form hydrogen bonds with water.

23. Acetone is an important organic solvent.

24. Camphor is used in foods as peppermint flavoring.

25. Cinnamon flavoring contains cinnamaldehyde.

ANSWERS TO PROGRAMMED REVIEW

13.1 a) hydrogen   b) -al   c) one   d) -one

13.2 a) below   b) hydrogen   c) above   d) soluble

13.3 a) carboxylic acid   b) oxidizing agents   c) Ag   d) Pt
      e) acetal

13.4 a) formaldehyde   b) acetone   c) aldehyde

ANSWERS TO SELF-TEST QUESTIONS

1.	a	10.	b	18.	b
2.	c	11.	c	19.	T
3.	c	12.	d	20.	F
4.	c	13.	b	21.	F
5.	a	14.	d	22.	T
6.	a	15.	c	23.	T
7.	c	16.	d	24.	F
8.	c	17.	a	25.	T
9.	a				

CHAPTER 14

Carboxylic Acids and Esters

PROGRAMMED REVIEW

Section 14.1 Nomenclature of Carboxylic Acids

The characteristic ending for carboxylic acid common names is (a) _____. The characteristic ending for carboxylic acid IUPAC names is (b) _____. Long-chain carboxylic acids derived from fats are referred to as (c) _____. The simplest aromatic carboxylic acid is (d) _____.

Section 14.2 Physical Properties of Carboxylic Acids

Carboxylic acid dimers contain (a) _____ molecules hydrogen bonded together. Carboxylic acids have (b) _____ boiling points than those of alcohols with comparable molecular weights. In comparing water solubility, carboxylic acids are (c) _____ soluble than aldehydes of comparable molecular weight. Low-molecular-weight carboxylic acids have sharp or (d) _____ odors.

Section 14.3 The Acidity of Carboxylic Acids

The negative ion formed as a carboxylic acid dissociates in water is the (a) _____ ion. In terms of acid strength, carboxylic acids are (b) _____ acids. At body pH carboxylic acids exist primarily as (c) _____ ions. Sodium hydroxide reacts with a carboxylic acid to produce a (d) _____ plus water.

Section 14.4 Salts of Carboxylic Acids

The characteristic ending for the names of salts of carboxylic acids is (a) _____. In terms of water solubility, carboxylate salts are (b) _____ soluble than the corresponding carboxylic acids. Sodium salts of long-chain carboxylic acids are useful as (c) _____. Certain carboxylic acid salts are used in foods as (d) _____ to prevent the growth of mold.

200 CHAPTER 14

Section 14.5 Carboxylate Esters

Carboxylic esters are formed from carboxylic acids and (a) _____. The process of ester formation is called (b) _____. The single bond between a carbonyl carbon and an oxygen in an ester is called the (c) _____ linkage.

Section 14.6 Nomenclature of Esters

An ester is comprised of a carboxylic acid portion and an alcohol component. The first word in naming an ester is derived from the (a) _____ portion of the structure. The IUPAC ending for an ester name is (b) _____. An ester formed from acetic acid would be named such that the second word of the name is (c) _____. An ester derived from ethyl alcohol and benzoic acid would be called (d) _____.

Section 14.7 Reactions of Esters

Ester hydrolysis may be catalyzed by (a) _____. The products of ester hydrolysis are a carboxylic acid and an (b) _____. The basic cleavage of an ester is termed (c) _____. The products of basic cleavage of an ester are an alcohol and a (d) _____.

Section 14.8 Esters of Inorganic Acids

Reaction of an alcohol with phosphoric acid can produce a (a) _____ ester. Phosphoric acid in which two hydrogens have been replaced by R groups may be referred to as a (b) _____. The letters ATP stand for adenosine (c) _____.

SOLUTIONS TO EXERCISES ANSWERED IN THE TEXT

14.1   An alcohol contains an —OH group, and a ketone has a C=O. A carboxyl group has both of these features.

14.3   acetic acid or ethanoic acid (IUPAC name)

14.5   a) pentanoic acid
       c) 4-phenylhexanoic acid
       e) 3-ethyl-5-methylhexanoic acid

14.6 a) H–C(=O)–OH   c) CH₃CH₂CH₂CH(CH₂CH₃)CH₂–C(=O)–OH

14.8 a) Acetic acid is higher boiling because it can form hydrogen-bonded dimers.
c) Butyric acid is higher boiling because it has the higher molecular weight of the two carboxylic acids.

14.9 c, a, d, b

14.11 b) Propanoic acid is more water-soluble because it can hydrogen bond more extensively with water.

14.13 Sodium acetate and sodium caprate are water-soluble because of their ionic character. Acetic acid is water-soluble due to hydrogen bonding with the carboxyl group. Capric acid is insoluble because of the extensive nonpolar hydrocarbon portion of the molecule.

14.14 b, d, a, c

14.16 Under cellular pH conditions, lactic acid ionizes to produce the anion lactate.

14.18 a) CH₃CH₂–C(=O)–O⁻Na⁺ + H₂O   c) CH₃(CH₂)₇–C(=O)–O⁻Na⁺ + H₂O

14.19 a) CH₃–C(=O)–OH + NaOH → CH₃–C(=O)–O⁻Na⁺ + H₂O

c) 2CH₃–C(=O)–OH + Ca(OH)₂ → (CH₃–C(=O)–O⁻)₂Ca²⁺ + 2H₂O

14.20 a) sodium propanoate   c) sodium *m*-nitrobenzoate

14.21 a) H–C(=O)–O⁻Na⁺   c) (C₆H₅–C(=O)–O⁻)₂ Ca²⁺

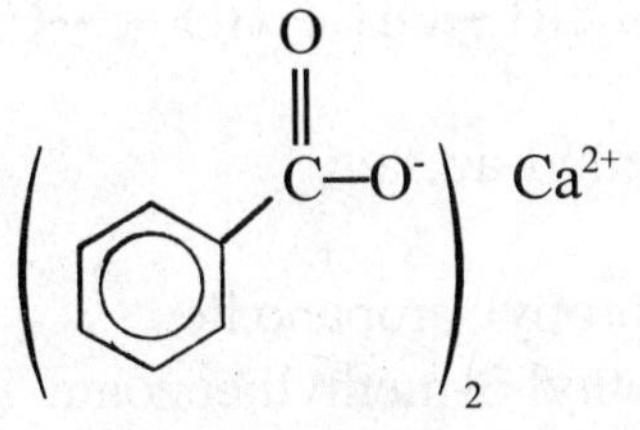

14.22 b) magnesium lactate

14.23 a) sodium stearate   e) sodium propionate
c) sodium benzoate

14.24 b, d, f

14.25 b) 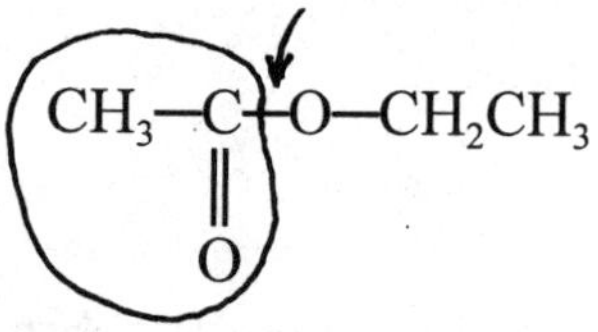 CH₃—C(=O)—O—CH₂CH₃ (bond between C and O circled, arrow at O)

d) C₆H₅—O—C(=O)—CH₃ (bond between O and C circled, arrow)

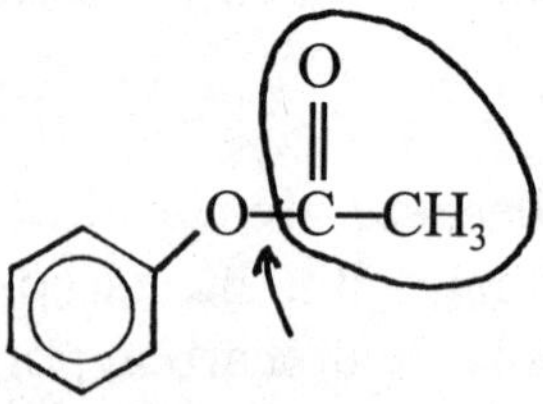

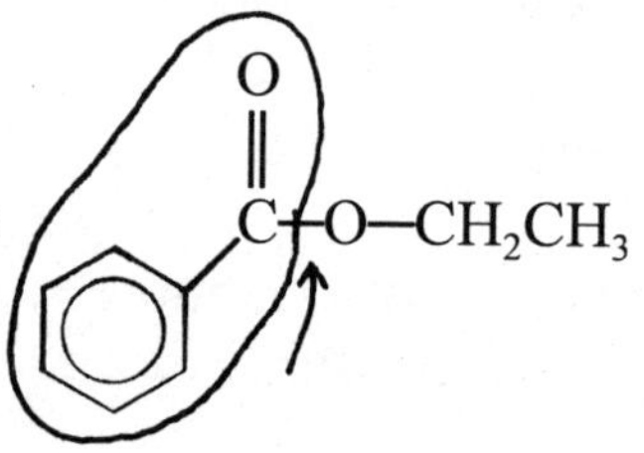

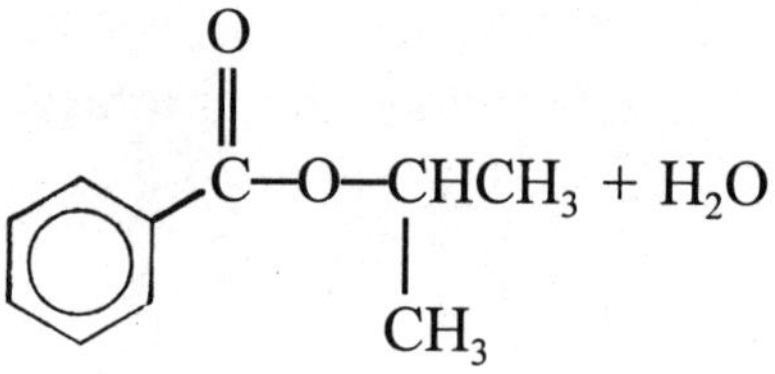

f) C₆H₅—C(=O)—O—CH₂CH₃ (C=O and phenyl circled, arrow)

14.26 a) C₆H₅—C(=O)—O—CH(CH₃)CH₃ + H₂O

e) 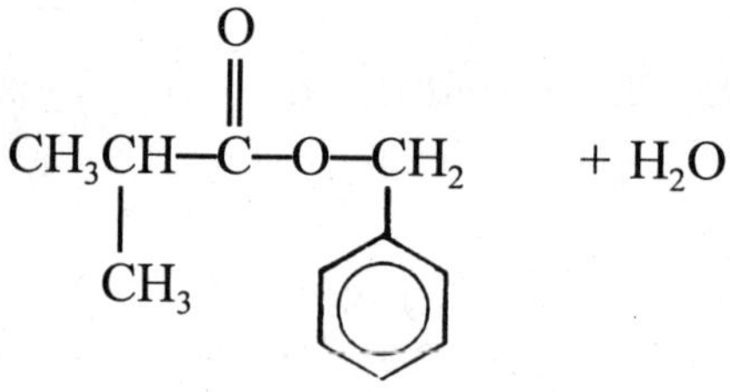 CH₃CH(CH₃)—C(=O)—O—CH₂—C₆H₅ + H₂O

c) CH₃—C(=O)—O—CH₃ + H₂O

14.28 b) 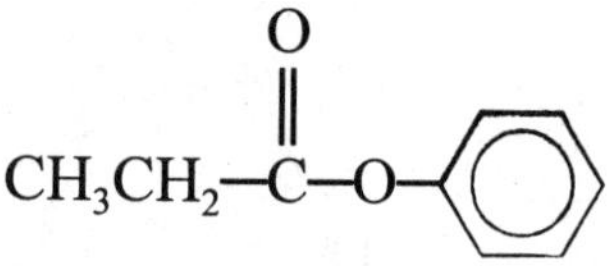 CH₃CH₂—C(=O)—O—C₆H₅

14.29 a) CH₃CH₂—OH + HO—C(=O)—CH₂CH₂CH₃

14.31 a) methyl acetate   c) ethyl lactate

14.32 a) n-propyl propanoate   e) ethyl methanoate
c) methyl 3-methylbenzoate

14.33 b) *n*-propyl ethanoate

14.34 a) methyl propionate c) methyl lactate

14.35 a) $CH_3CH_2CH_2-\underset{\underset{O}{\|}}{C}-OCH_2CH_3$

c) $CH_3CH_2CH_2CH_2-\underset{\underset{O}{\|}}{C}-OCH_2CH_2CH_3$

e) $O_2N-\underset{}{C_6H_4}-\underset{\underset{O}{\|}}{C}-OCH_3$

14.37 $CH_3-\underset{\underset{O}{\|}}{C}-OCH_3 + H_2O \xrightarrow{H^+} CH_3-\underset{\underset{O}{\|}}{C}-OH + CH_3-OH$

$CH_3-\underset{\underset{O}{\|}}{C}-OCH_3 + NaOH \rightarrow CH_3-\underset{\underset{O}{\|}}{C}-O^-Na^+ + CH_3-OH$

14.38 a) $CH_3\underset{\underset{CH_3}{|}}{CH}-\underset{\underset{O}{\|}}{C}-OH + CH_3CH_2-OH$

c) $CH_3(CH_2)_{16}-\underset{\underset{O}{\|}}{C}-O^-Na^+ + CH_3CH_2-OH$

e) $HO-\underset{\underset{O}{\|}}{C}-CH_2-\underset{\underset{O}{\|}}{C}-OH + 2CH_3-OH$

14.39 b) Under cellular pH conditions, the anion is favored over the acidic form.

14.41 a) 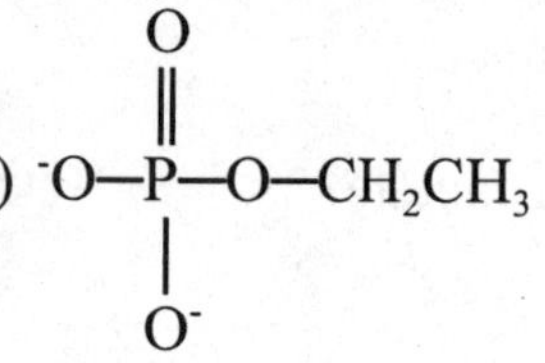 ⁻O—P(=O)(O⁻)—O—CH₂CH₃

c) 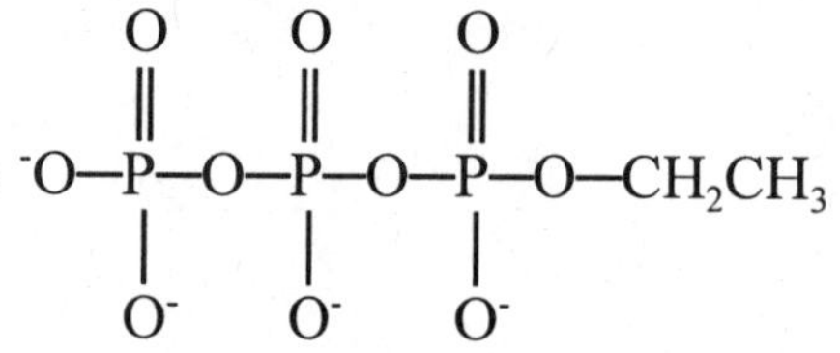 ⁻O—P(=O)(O⁻)—O—P(=O)(O⁻)—O—P(=O)(O⁻)—O—CH₂CH₃

14.42 a) fever reducer b) dacron fibers

14.44 Polyester is wrinkle-resistant. Polyester is blended with cotton to give a fabric that is comfortable to wear on hot, humid days.

14.46 —(—O—C(=O)—C(=O)—O—CH₂CH₂CH₂—)ₙ—

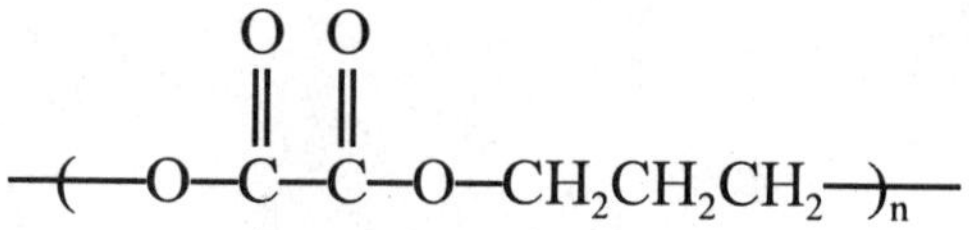

14.47 tricarboxylic

14.49 The reaction is carried out under basic conditions where any carboxylic acid formed would be converted to the carboxylate salt.

SELF-TEST QUESTIONS

Multiple Choice

1. A correct IUPAC name for (2-bromobenzoic acid structure) is

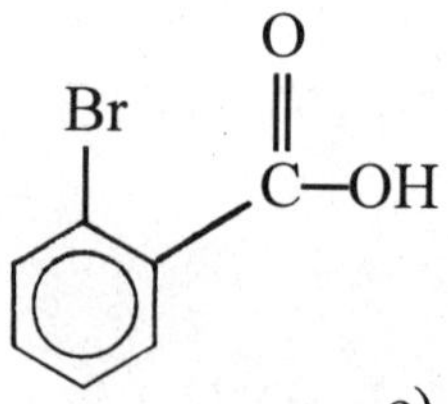

   a) bromobenzoic acid
   b) 1-bromobenzoic acid
   c) 2-bromobenzoic acid
   d) 2-bromo-1-benzoic acid

2. A correct IUPAC name for CH₃CHCH₂COOH is
                              |
                             CH₃

   a) 2-methylpentanoic acid   c) 3-methylbutanoic acid
   b) 2-methylbutanoic acid   d) 3-methylpentanoic acid

3. Which of the following pure substances would exhibit hydrogen bonding?
   a) aldehyde   c) ether
   b) ketone   d) carboxylic acid

4. Which of the following substances would you expect to be the most soluble in water?

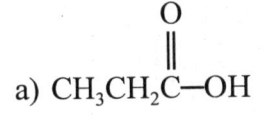

   a) CH₃CH₂C(=O)—OH   c) CH₃C(=O)—OCH₃

   b) CH₃C(=O)—OH   d) CH₃C(=O)—OCH₂CH₃

5. Which name is more appropriate for the organic acid under body conditions of pH 7.4?
   a) lactic acid   b) lactate

6. The reaction of butanoic acid and NaOH produces
   a) an ester   c) a carboxylic acid and an alcohol
   b) a ketone   d) a carboxylate salt

7. What reagent is needed to complete the following reaction?

   Ph—COOH + ? → Ph—COO⁻K⁺ + H₂O

   a) K   b) K⁺   c) KOH   d) KO₂

8. Which of the following molecules could be used as one of the reagents necessary to prepare

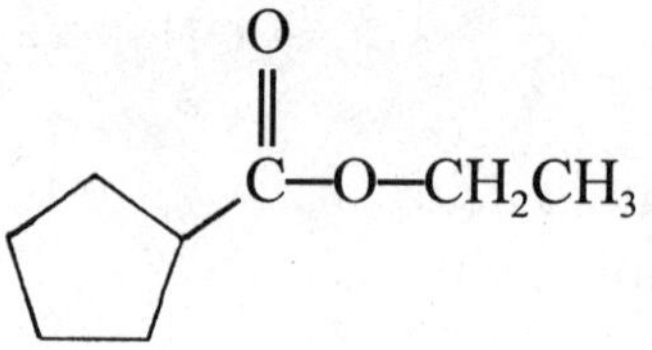

[cyclopentyl-C(=O)-O-CH₂CH₃]

a) CH₃CH₂OH   b) CH₃C(=O)-OH   c) 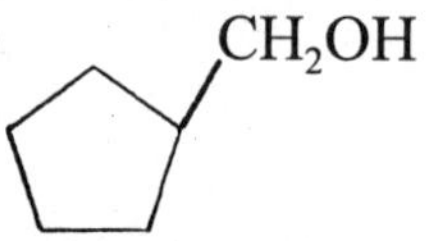 [cyclopentyl-CH₂OH]   d) 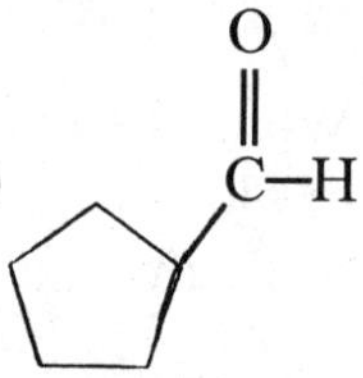 [cyclopentyl-C(=O)H]

9. The organic product of the following reaction is

$$CH_3C(=O)-OH + CH_3-OH \xrightarrow{H^+}$$

a) CH₃C(=O)-CH₂-OH
b) CH₃C(=O)-O-CH₂-OH
c) CH₃C(=O)-CH₃
d) CH₃C(=O)-OCH₃

10. The ester formed by reacting propanoic acid and isopropyl alcohol is
   a) propyl propanoate
   b) isopropyl propanoic acid
   c) isopropyl propanoate
   d) 2-propyl propanoate

11. The IUPAC name for the ester formed in the reaction of isopropyl alcohol and benzoic acid is
   a) benzyl isopropyl ester
   b) benzyl isopropanoate
   c) isopropyl benzoate
   d) isopropyl benzoic acid

12. Which of the following materials might be obtained as one of the products from the reaction

$$CH_3C(=O)-O-CH_2CH_2CH_3 + NaOH \rightarrow \underline{\qquad}$$

a) CH₃CH₂OH   b) CH₃CH₂C(=O)-O⁻Na⁺   c) CH₃C(=O)-O⁻Na⁺   d) CH₃C(=O)-OH

Matching

Select the best match for each of the following:

13. CH₃C(=O)—OH

14. 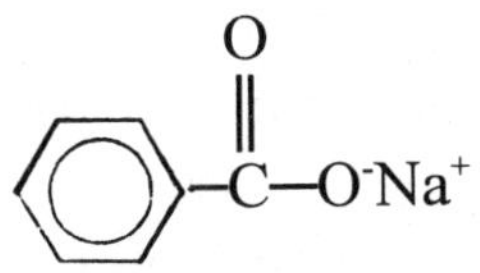 C₆H₅—C(=O)—O⁻Na⁺

15. CH₃(CH₂)₁₆C(=O)—O⁻Na⁺

16. CH₃C(=O)—OCH₂CH₃

a) a preservative used in pop
b) a soap
c) present in vinegar
d) fingernail polish remover

For each reaction on the left, choose the correct description from the responses on the right.

17. ester + H₂O $\xrightarrow{H^+}$

18. ester + NaOH →

19. carboxylic acid + H₂O →

20. carboxylic acid + alcohol $\xrightarrow[\text{heat}]{H^+}$

a) dissociation
b) esterification
c) hydrolysis
d) saponification

True-False

21. The boiling points of carboxylic acids are lower than those of the corresponding alcohols.

22. Salts of carboxylic acids are not usually soluble in water.

23. Hydrogen bonding increases both the boiling points and the water solubility of carboxylic acids.

24. CH$_3$COOH has a higher boiling point than CH$_3$CH$\overset{\overset{\displaystyle O}{\|}}{{}}$.

25. Carboxylic acids are generally strong acids.

26. Both nitric acid and phosphoric acid can react with alcohols to form esters.

27. Certain phosphate esters are present in the body.

ANSWERS TO PROGRAMMED REVIEW

14.1 a) -ic acid   b) -oic acid   c) fatty acids   d) benzoic acid

14.2 a) two   b) higher   c) more   d) unpleasant

14.3 a) carboxylate   b) weak   c) carboxylate   d) salt

14.4 a) -ate   b) more   c) soaps   d) preservatives

14.5 a) alcohols   b) esterification   c) ester

14.6 a) alcohol   b) -ate   c) acetate   d) ethyl benzoate

14.7 a) H$^+$ b) alcohol   c) saponification   d) salt

14.8 a) phosphate   b) diester   c) triphosphate

ANSWERS TO SELF-TEST QUESTIONS

1.	c	10.	c	19.	a
2.	c	11.	c	20.	b
3.	d	12.	c	21.	F
4.	b	13.	c	22.	F
5.	b	14.	a	23.	T
6.	d	15.	b	24.	T
7.	c	16.	d	25.	F
8.	a	17.	c	26.	T
9.	d	18.	d	27.	T

CHAPTER 15

Amines and Amides

PROGRAMMED REVIEW

Section 15.1 Classification of Amines

An amine having one alkyl or aromatic group bonded to nitrogen is classified as a (a) _____ amine. A tertiary amine has (b) _____ alkyl or aromatic groups bonded to nitrogen. The amine CH_3NHCH_3 is classified as a (c) _____ amine.

Section 15.2 Nomenclature of Amines

In the IUPAC system, the $-NH_2$ group is called the (b) _____ group. The IUPAC name for ethylamine is (b) _____. The common name for $CH_3CH_2CH_2-NH-CH_3$ is (c) _____. Aromatic amines are named as derivatives of (d) _____.

Section 15.3 Physical Properties of Amines

(a) _____ amines do not form hydrogen bonds among themselves. A primary amine has a boiling point somewhat (b) _____ than the boiling point of an alcohol with similar molecular weight. Tertiary amines have boiling points similar to those of (c) _____ with similar molecular weights. In terms of water solubility, amines with fewer than six carbons are generally (d) _____.

Section 15.4 Chemical Properties of Amines

The most distinguishing feature of amines is their behavior as weak (a) _____. Methylamine reacts with water to form $CH_3-NH_3^+$ and (b) _____. Amines react with acids to form (c) _____. Amine salts tend to be (d) _____ soluble in water than the parent amines. Amine salts in which (e) _____ alkyl groups are bonded to nitrogen are called quaternary ammonium salts. A primary amine can react with an acid chloride to form an (f) _____.

211

Section 15.5 Biologically Important Amines

A (a) _____ acts as a chemical bridge in nerve impulse transmission between nerve cells. Another name for adrenalin, the "fight or flight" hormone, is (b) _____. A class of drugs structurally similar to adrenalin which are stimulants of the central nervous system is (c) _____. Because alkaloids are (d) _____-containing, they are usually weak bases. An alkaloid present in tobacco is (e) _____. The biological source of codeine is the (f) _____ plant. Heroin is a chemical derivative of (g) _____.

Section 15.6 Amide Nomenclature

The characteristic IUPAC ending for amide names is (a) _____. The IUPAC name for
$$CH_3CH_2\overset{O}{\overset{\|}{C}}-NH_2$$
is (b) _____. In the name N-methylbenzamide, the letter N signifies the methyl group is attached to (c) _____.

Section 15.7 Physical Properties of Amides

Low-molecular-weight amides are water soluble because of their ability to form (a) _____ bonds with water. Most amides exist as solids because of their ability to form a network of intermolecular (b) _____ bonds. (c) _____ amides often have lower boiling points than monosubstituted amides.

Section 15.8 Chemical Properties of Amides

Hydrolysis of amides under acidic conditions produces a (a) _____ and an (b) _____ salt. Hydrolysis of amides under basic conditions produces an (c) _____ and a (d) _____ salt.

SOLUTIONS TO EXERCISES ANSWERED IN THE TEXT

15.2 a) secondary   e) secondary
      c) primary

15.4 a) diethylamine   c) dimethyl-*n*-propylamine

15.5 a) 2-amino-4-methylpentane   e) 4-amino-2-pentene
      c) 1,2-diaminocyclopentane

15.6 a) 2,6-dimethylaniline c) *N,N*-dimethylaniline

15.7 a) *t*-butylamine or 2-amino-2-methylpropane
 c) 3-chloroaniline or *m*-chloroaniline

15.8 a) CH$_3$CH(NH$_2$)CH$_2$CH(CH$_3$)CH$_2$CH$_2$CH$_3$

 e) CH$_3$CH$_2$–N(CH$_3$)–CH$_2$CH$_3$

 c) (C$_6$H$_5$)$_3$N (triphenylamine)

15.9 Amines form hydrogen bonds with water.

15.11 Tertiary amines do not form hydrogen bonds among themselves.

15.12 a) CH$_3$CH$_2$–N(H)–CH$_3$ ⋯ H–N(CH$_3$)–CH$_2$CH$_3$

 c) H–O–H ⋯ N(CH$_3$)$_3$

15.14 CH$_3$CH$_2$–NH–CH$_2$CH$_3$ + H$_2$O ⇌ CH$_3$CH$_2$–N$^+$H$_2$–CH$_2$CH$_3$ + OH$^-$

15.15 a) $CH_3CH_2CH_2CH_2-NH_3{}^+Cl^-$

c) cyclopentyl-$NH_3{}^+CH_3C(=O)-O^-$

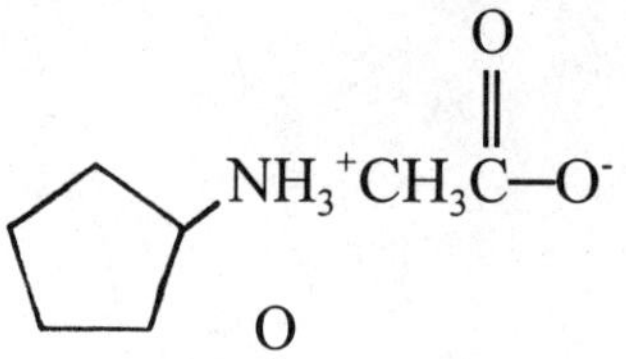

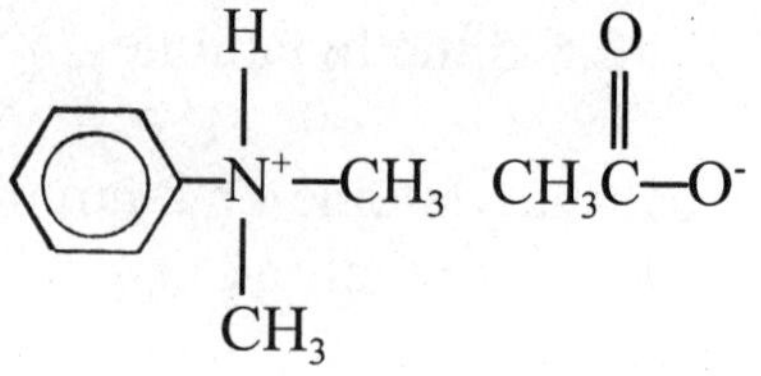

e) $CH_3CH(CH_3)-C(=O)-NH-CH_3 + H_2O$

g) $C_6H_5-N^+(H)(CH_3)(CH_3)$ $CH_3C(=O)-O^-$

i) no reaction

15.16 The N atom in a quaternary salt is bonded to four carbon atoms. In the salt of a tertiary amine, the N atom is bonded to three carbon atoms.

15.18 [structure] + HCl → [protonated structure with Cl⁻]

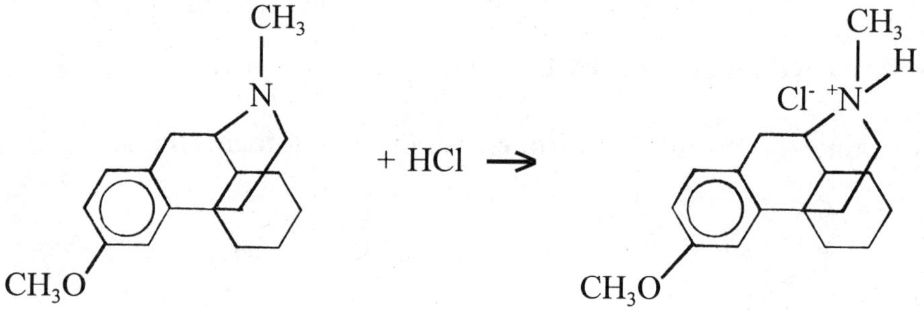

15.20 a) $C_6H_5-NH_3{}^+Cl^- + NaOH$

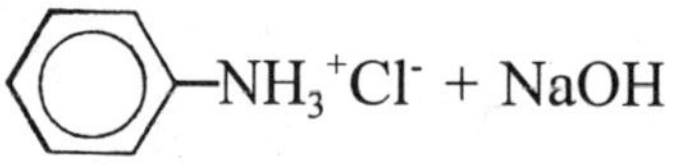

c) $CH_3CH_2-NH_2 + H_2SO_4$

e) $CH_3CH(CH_3)-NH_2 + H_2O$

15.21 No additional products for parts a, c, or e.

15.22 Neurotransmitters serve as chemical bridges in nerve impulse transmission across the nerve synapses.

15.24 Epinephrine is released in emergency situations to help furnish the body tissues with a supply of glucose.

15.26 Amphetamines are powerful stimulants of the nervous system. Abusers of amphetamines experience sleeplessness, weight loss, and paranoia.

15.28 They contain nitrogen atoms.

15.29 a) caffeine
c) nicotine
e) quinine

15.30 a) propanamide   e) 2-methylbutanamide
       c) N-ethylbenzamide

15.31 a) $CH_3CH_2CH_2-\overset{\overset{O}{\|}}{C}-NH_2$

e) $H-\overset{\overset{O}{\|}}{C}-\underset{\underset{CH_3}{|}}{N}-CH_3$

c) $CH_3-\underset{}{\bigcirc}-\overset{\overset{O}{\|}}{C}-NH_2$

15.32 a) $CH_3-\overset{\overset{O\cdots H\diagdown\;\diagup H}{\|}\;\;\;\;\;O}{\underset{\underset{CH_3}{|}}{C}-N-CH_3}$

c) $CH_3-\overset{\overset{O}{\|}}{C}-\underset{\underset{H}{|}}{N}-H\cdots\underset{\underset{H}{|}}{O}-H$

15.34 a) $CH_3-\overset{\overset{O}{\|}}{C}-O^-Na^+ + CH_3CH_2-NH_2$

c) $\bigcirc-\overset{\overset{O}{\|}}{C}-O^-Na^+ + CH_3-NH-CH_3$

15.36 [2,6-dimethylphenyl-$NH_3^+Cl^-$] + $HO-\overset{\overset{O}{\|}}{C}-CH_2\underset{\underset{CH_2CH_3}{|}}{\overset{\overset{H}{|}}{N^+}}CH_2CH_3$

15.38 Some antihistamines produce drowsiness.

15.40 many amide groups

15.42 Abuse of cocaine may cause chronic depression. An overdose may cause death through cardiac or respiratory arrest.

15.44 Reye's syndrome has been correlated with the use of aspirin by children who are feverish with the flu or chicken pox.

15.46 Ibuprofen may be superior to aspirin as an antiinflammatory drug and is not correlated with Reye's syndrome.

SELF-TEST QUESTIONS

Multiple Choice

1. The amine CH₃CH—NH₂ with CH₃ substituent is classified as

   a) primary
   b) secondary
   c) tertiary
   d) quaternary

2. Which of the following is a secondary amine?

   a) CH₃CH₂CHCH₃ with NH₂ substituent
   b) CH₃—N—CH₂CH₃ with CH₃ substituent
   c) CH₃CH₂—NH₂
   d) 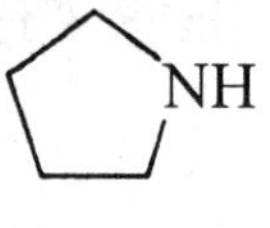 pyrrolidine ring with NH

3. A common name for CH₃CH₂—NH—CH₂CH₃ is

   a) ethylaminoethane
   b) diethylamine
   c) aminoethylethane
   d) diethylammonia

4. The IUPAC name for CH₃CHCH₂CHCH₃ (with NH₂ and CH₃ substituents) is

   a) 4-amino-2-methylpentane
   b) 4-amino-2-isopropylpropane
   c) 2-amino-4-methylpentane
   d) 2-amino-1-isopropylpropane

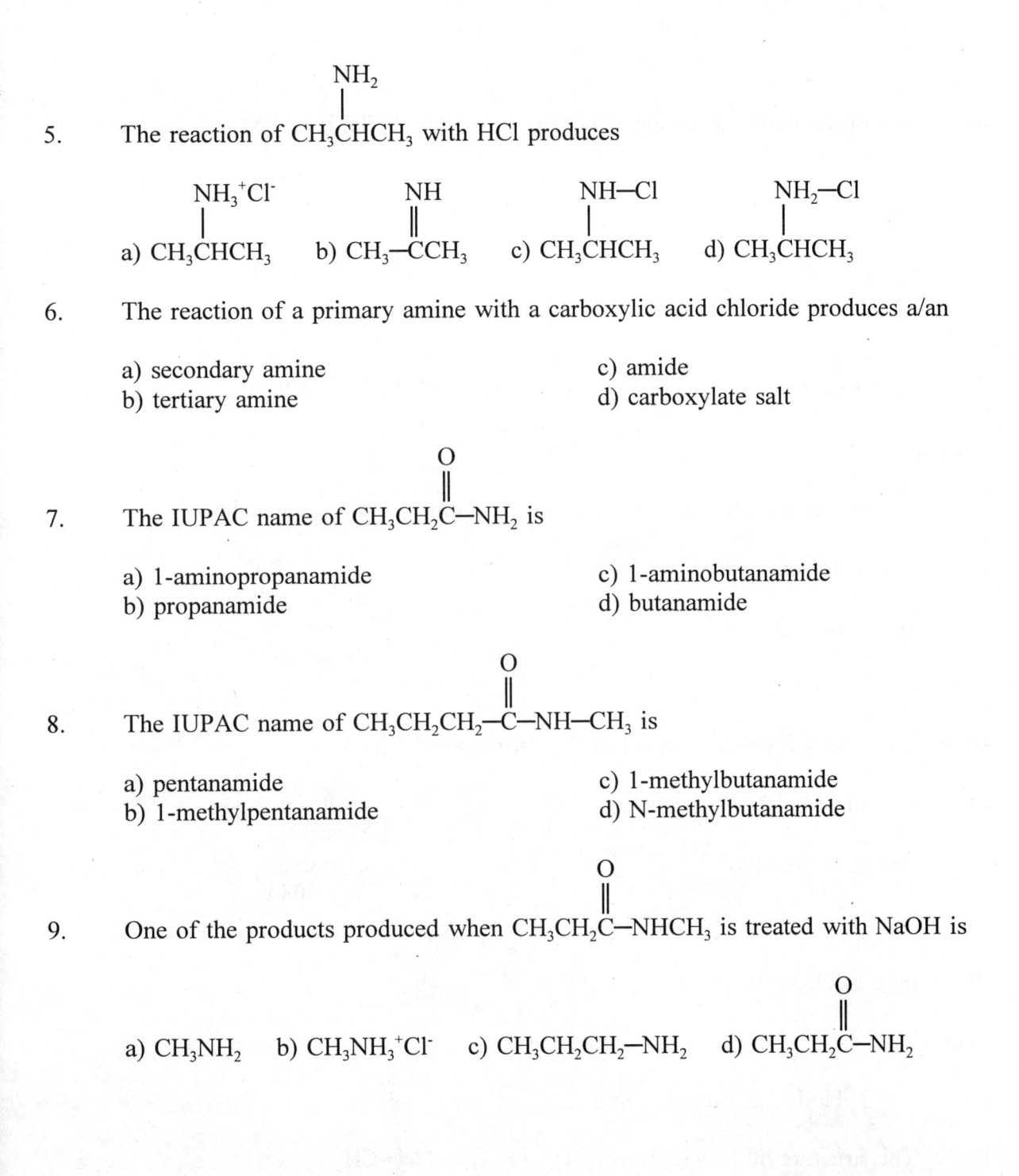

5. The reaction of CH$_3$CH(NH$_2$)CH$_3$ with HCl produces

a) CH$_3$CH(NH$_3^+$Cl$^-$)CH$_3$
b) CH$_3$—C(=NH)CH$_3$
c) CH$_3$CH(NH—Cl)CH$_3$
d) CH$_3$CH(NH$_2$—Cl)CH$_3$

6. The reaction of a primary amine with a carboxylic acid chloride produces a/an

a) secondary amine
b) tertiary amine
c) amide
d) carboxylate salt

7. The IUPAC name of CH$_3$CH$_2$C(=O)—NH$_2$ is

a) 1-aminopropanamide
b) propanamide
c) 1-aminobutanamide
d) butanamide

8. The IUPAC name of CH$_3$CH$_2$CH$_2$—C(=O)—NH—CH$_3$ is

a) pentanamide
b) 1-methylpentanamide
c) 1-methylbutanamide
d) N-methylbutanamide

9. One of the products produced when CH$_3$CH$_2$C(=O)—NHCH$_3$ is treated with NaOH is

a) CH$_3$NH$_2$
b) CH$_3$NH$_3^+$Cl$^-$
c) CH$_3$CH$_2$CH$_2$—NH$_2$
d) CH$_3$CH$_2$C(=O)—NH$_2$

218 CHAPTER 15

10. One of the products produced when CH₃C(=O)—NHCH₂CH₃ is treated with HCl and H₂O is

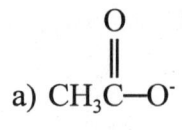

a) CH₃C(=O)—O⁻

b) CH₃C(=O)—NH₂

c) CH₃CH₂—NH₃⁺Cl⁻

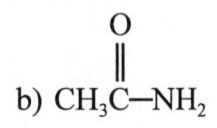

d) CH₃CH₂—NH₂

Matching

For each description on the left, select an amide from the list on the right.

11. a tranquilizer

12. an aspirin substitute

13. an antibiotic

a) polycillin
b) benzamide
c) acetaminophen
d) valium

For each description on the left, select the correct alkaloid.

14. present in coffee

15. present in tobacco

16. used as a cough suppressant

17. used to dilate the pupil of the eye

a) atropine
b) nicotine
c) codeine
d) caffeine

True-False

18. Triethylamine is a tertiary amine.

19. The structure of 1-methylaniline is C₆H₅—NH—CH₃ .

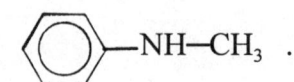

20. Tertiary amines have higher boiling points than primary and secondary amines.

21. Low-molecular-weight amines have a characteristic pleasant odor.

22. Both primary and tertiary amines can hydrogen bond with water molecules.

23. Amines under basic conditions exist as amine salts.

24. Disubstituted amides have lower boiling points than unsubstituted amides.

25. Amides with fewer than six carbons are water soluble.

26. Amide molecules are neither basic nor acidic.

27. Serotonin is an important neurotransmitter.

28. Amphetamine is a powerful nervous system stimulant.

ANSWERS TO PROGRAMMED REVIEW

15.1 a) primary b) three c) secondary

15.2 a) amino b) aminoethane c) methyl-*n*-propylamine
      d) aniline

15.3 a) tertiary b) lower c) alkanes d) soluble

15.4 a) bases b) OH⁻ c) salts d) more d) four
      f) amide

15.5 a) neurotransmitter b) epinephrine c) amphetamines
      d) nitrogen e) nicotine f) poppy g) morphine

15.6 a) -amide b) propanamide c) nitrogen

15.7 a) hydrogen b) hydrogen c) disubstituted

15.8 a) carboxylic acid b) amine c) amine d) carboxylic acid

ANSWERS TO SELF-TEST QUESTIONS

1. a
2. d
3. b
4. c
5. a
6. c
7. b
8. d
9. a
10. c
11. d
12. c
13. a
14. d
15. b
16. c
17. a
18. T
19. F
20. F
21. F
22. T
23. F
24. T
25. T
26. T
27. T
28. T

CHAPTER 16

Carbohydrates

PROGRAMMED REVIEW

Section 16.1 Classes of Carbohydrates

Carbohydrates may be defined as (a) _____ aldehydes or ketones or substances that yield such compounds upon hydrolysis. Carbohydrates formed by the combination of two monosaccharides are known as (b) _____. Carbohydrates formed by the combination of many monosaccharide units are known as (c) _____.

Section 16.2 Stereochemistry of Carbohydrates

Isomers which are mirror images of each other are called (b) _____. A chiral carbon atom has (b) _____ different groups attached. The number of stereoisomers possible in a structure containing n chiral carbon atoms is (c) _____.

Section 16.3 Fischer Projections

In a Fischer projection a chiral carbon is represented by the (a) _____ of two lines. The two bonds coming toward the viewer in a Fischer projection are drawn (b) _____. A small capital D is used to indicate that an -OH group is on the (c) _____ in a Fischer projection. The (d) _____ enantiomer rotates a plane of polarized light to the left. The (e) _____ enantiomers of monosaccharides are preferred by the human body.

Section 16.4 Monosaccharides

Sugars with four carbon atoms are known as (a) _____. The presence of an aldehyde group in a monosaccharide may be designated by the use of the prefix (b) _____. Of the eight aldopentoses (c) _____ belong to the D series. A ketose is a carbohydrate containing a (d) _____ group.

221

Section 16.5 Physical Properties of Monosaccharides

Monosaccharides and disaccharides are also called (a) _____ because they taste sweet. Monosaccharides are very soluble in water because they contain several (b) _____ groups which hydrogen bond with water.

Section 16.6 Chemical Properties of Monosaccharides

The ring formed as glucose undergoes cyclization is referred to as a (a) _____ ring. The cyclization of fructose results in the formation of a (b) _____ ring. Glucose with an OH group pointing up at position 1 is called (c) _____. Sugars that can be oxidized by weak oxidizing agents are called (d) _____. A name for carbohydrates containing an acetal group is (e) _____.

Section 16.7 Important Monosaccharides

The monosaccharide component of deoxyribonucleic acid is (a) _____. The common monosaccharide with an OH group directed up at position 4 is (b) _____. The monosaccharide known as blood sugar is (c) _____. The sweetest of the common sugars is (d) _____.

Section 16.8 Disaccharides

A disaccharide containing an α (1 → 4) linkage is (a) _____. A disaccharide containing a fructose component is (b) _____. The disaccharide (c) _____ is sometimes referred to as milk sugar. The hydrolysis of sucrose produces a mixture referred to as (d) _____ sugar.

Section 16.9 Polysaccharides

Starch has a linear form called (a) _____ and a branched form called (b) _____. Glycogen has both α (1 → 4) and (c) _____ linkages. Cellulose is a linear polymer of glucose units linked (d) _____.

Carbohydrates

SOLUTIONS TO EXERCISES ANSWERED IN THE TEXT

16.1 They provide energy through their oxidation, carbon for the synthesis of cell components, a stored form of chemical energy, and a part of the structural elements of some cells and tissues.

16.2   b) an energy source in our diet
     c) a stored form of energy in animals

16.4 They differ in the number of polyhydroxy aldehyde or ketone units present.

16.5   b) monosaccharide, carbohydrate   f) disaccharide, carbohydrate
     d) monosaccharide, carbohydrate   h) polysaccharide, carbohydrate

16.7

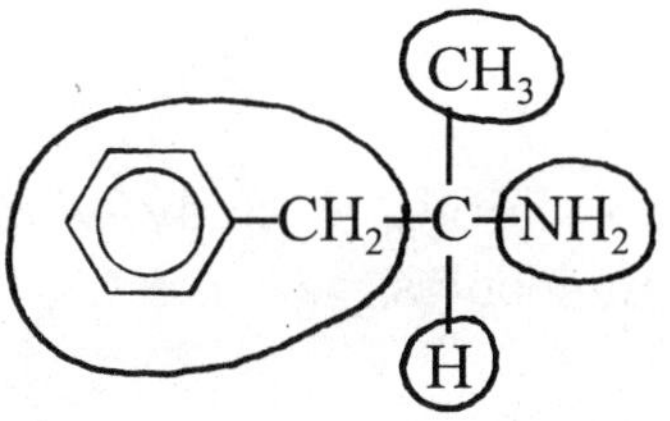

16.8   b)

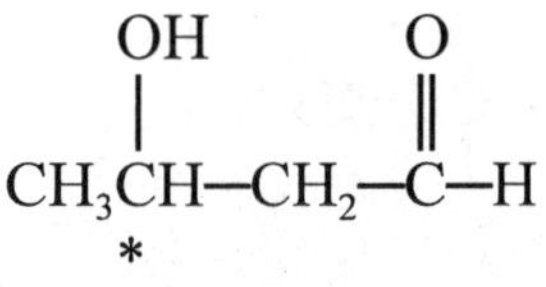

     f)

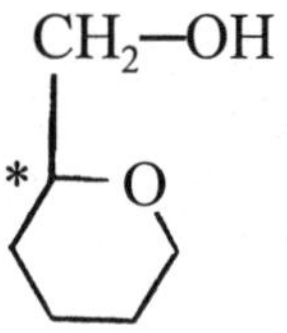

     d) no chiral carbon atoms

16.10   b) the D form is given in the text   d) the L form is given in the text.

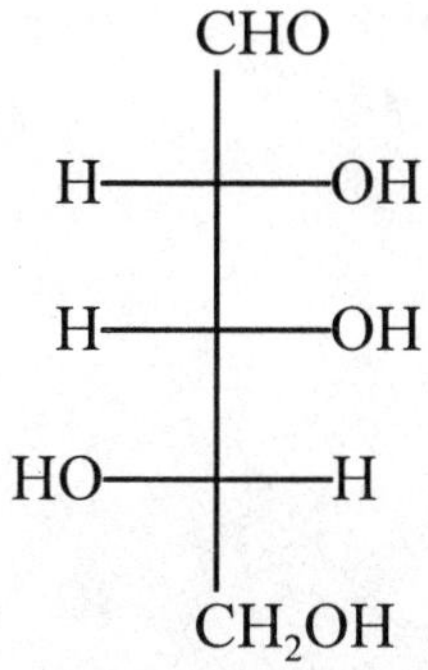

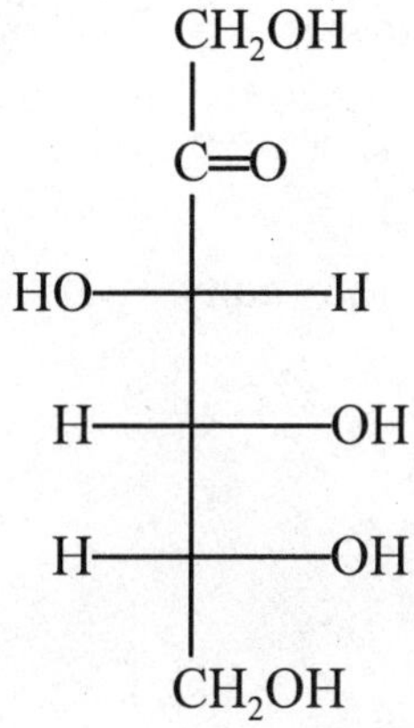

enantiomer (L)   enantiomer (D)

224 CHAPTER 16

16.11 b)

```
         COOH                    COOH
          |                       |
   H —————+————— NH₂       H₂N —————+————— H
          |                       |
         CH₂OH                   CH₂OH

         D isomer                L isomer
```

16.12 b) Three chiral carbon atoms are marked with an asterisk.
Stereoisomers = $2^n = 2^3 = 8$.

```
      OH  OH  OH  OH
      |   |   |   |
     CH₂—CH—CH—CH—CHO
          *   *   *
```

16.14 Many of the compounds involved in biochemistry are chiral. A study of chirality helps to more fully understand the properties of biomolecules.

16.16 b) ketopentose   d) aldopentose

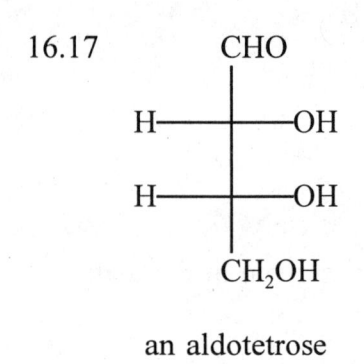

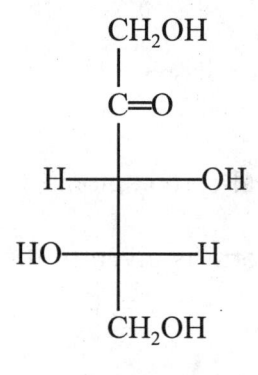

16.17
```
         CHO                    CH₂OH
          |                      |
   H —————+————— OH              C=O
          |                      |
   H —————+————— OH       H —————+————— OH
          |                      |
         CH₂OH             HO —————+————— H
                                  |
                                 CH₂OH

      an aldotetrose          a ketopentose
```

16.19 Because of the many —OH groups present

16.20 b) the β form is given in the text d) the β form is given in the text

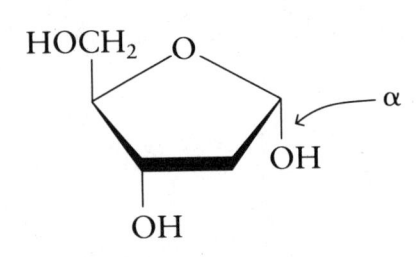

[structure: HOCH₂, O, OH, OH, α]   [structure: HOCH₂, O, CH₂OH, HO, OH, OH, α]

16.22 Pyranose rings contain six atoms, and furanose rings contain five atoms.

16.24 The cyclic structure is in equilibrium with the open-chain form that contains the reactive aldehyde group.

16.25 b)

[structure: CH₂OH, O, OH, HO, OCH₂CH₃, OH] + [structure: CH₂OH, O, OCH₂CH₃, OH, HO, OH] + H₂O

16.26 b)

[structure: HOCH₂, O, CH₂OH, HO, OCH₂CH₃, OH — glycosidic linkage]

16.28 Glucose is naturally present within the bloodstream and is used to supply energy to the tissues.

16.30 a) honey, grapes c) milk (present in lactose)
      b) honey, fruits

16.32 b) lactose f) sucrose
      d) maltose

16.34 acetal or ketal

16.35 b) maltose + water $\xrightarrow{\text{acid}}$ glucose + glucose

d) cellobiose + water $\xrightarrow{\text{acid}}$ glucose + glucose

16.36 because invert sugar contains fructose

16.38

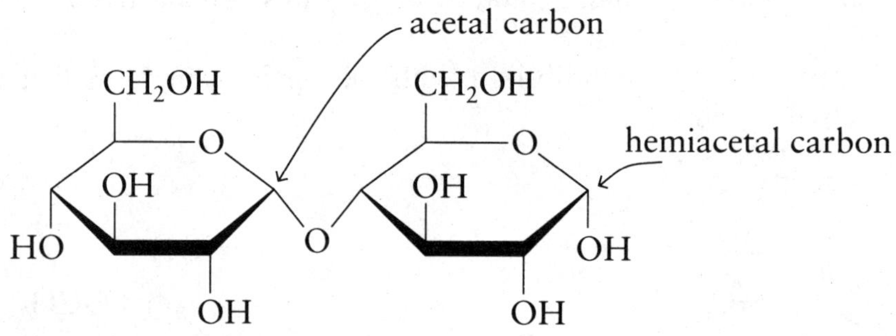

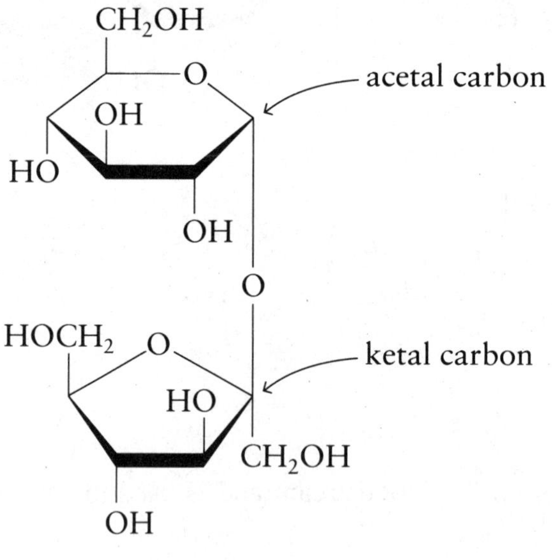

16.40 b) a reducing sugar because the ring on the right has a hemiacetal group

16.41 b) $\alpha(1 \rightarrow 4)$

16.43 a) galactose, glucose, fructose
     b) no, there is no hemiacetal or hemiketal group present
     c) α(1 → 6) and α - 1 → β - 2

16.44 b) cellulose   d) cellulose

16.45 b) all   f) amylose
     d) amylopectin and glycogen

16.47 saccharin, cyclamate, aspartame

16.49 sucrose

16.51 linen and cotton

16.53 it provides bulk within the digestive tract

SELF-TEST QUESTIONS

Multiple Choice

1. Which of the following is a monosaccharide?
   a) amylose   b) ribose   c) cellulose   d) lactose

2. Which of the following is a polysaccharide?
   a) amylose   b) lactose   c) maltose   d) galactose

3. How many chiral carbon atoms are in

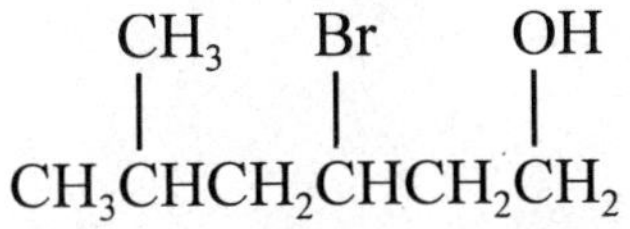

$$CH_3CHCH_2CHCH_2CH_2$$
with substituents CH₃, Br, OH

   a) 0   b) 1   c) 2   d) 3

4. How many chiral carbon atoms are in [structure: methylcyclohexane with Cl]

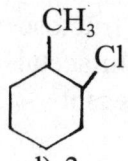

    a) 0 b) 1 c) 2 d) 3

5. How many stereoisomers are possible for

    $$\underset{\underset{\displaystyle CH_2}{|}}{OH}\;\underset{\underset{\displaystyle -CH}{|}}{OH}\;\underset{\underset{\displaystyle -CH-CHO}{|}}{OH}$$

    a) 0 b) 2 c) 4 d) 8

6. Glucose is a/an
   a) ketopentose
   b) ketohexose
   c) aldopentose
   d) aldohexose

7. Fructose is a/an
   a) ketopentose
   b) ketohexose
   c) aldopentose
   d) aldohexose

8. A positive Benedict's test is indicated by the formation of
   a) Ag b) CuOH c) Cu_2O d) Cu_2^+

Matching

For each monosaccharide described on the left, select the best response from the right.

9. given intravenously

10. present with glucose in invert sugar

11. Combines with glucose to form lactose

12. Found in genetic material

a) fructose
b) galactose
c) glucose
d) ribose

For each disaccharide described on the left, select the best match from the responses on the right.

13. used as household sugar

14. found in milk

15. formed in germinating grain

a) glycogen
b) sucrose
c) maltose
d) lactose

For each disaccharide on the left, select the correct hydrolysis products from the right.

16. sucrose

17. maltose

18. lactose

a) glucose and galactose
b) glucose and fructose
c) only glucose
d) hydrolysis does not occur

Select the correct polysaccharide for each description on the left.

19. a storage form of carbohydrates in animals

20. The most abundant polysaccharide in starch

21. primary constituent of paper

a) amylopectin
b) amylose
c) glycogen
d) cellulose

True-False

22. A D enantiomer is the mirror image of an L enantiomer.

23. In a D carbohydrate, the hydroxyl group on the bottom chiral carbon points to the left.

24. The L carbohydrates are preferred by the human body.

25. Sugars that contain a hemiacetal group are reducing sugars.

26. In β-galactose, the hydroxyl group at carbon 1 points up.

27. Maltose contains a glycosidic linkage.

28. The glucose ring of lactose can exist in an open-chain form.

29. Sucrose contains a hemiacetal group.

30. Linen is prepared from cellulose.

ANSWERS TO PROGRAMMED REVIEW

16.1 a) polyhydroxy   b) disaccharides   c) polysaccharides

16.2 a) enantiomers   b) four   c) 2^n

16.3 a) intersection   b) horizontally   c) right   d) levorotatory
      e) D

16.4 a) tetroses   b) aldo   c) four   d) ketone

16.5 a) sugars   b) OH

16.6 a) pyranose   b) furanose   c) β-glucose   d) reducing sugars
      e) glycoside

16.7 a) deoxyribose   b) galactose   c) glucose   d) fructose

16.8 a) maltose   b) sucrose   c) lactose   d) invert

16.9 a) amylose   b) amylopectin   c) α(1 → 6)   d) β(1 → 4)

ANSWERS TO SELF-TEST QUESTIONS

1.	b	11.	b	21.	d
2.	a	12.	d	22.	T
3.	b	13.	b	23.	F
4.	c	14.	d	24.	F
5.	c	15.	c	25.	T
6.	d	16.	b	26.	T
7.	b	17.	c	27.	T
8.	c	18.	a	28.	T
9.	c	19.	c	29.	F
10.	a	20.	a	30.	T

CHAPTER 17

Lipids

PROGRAMMED REVIEW

Section 17.1 Classification of Lipids

Saponifiable lipids all contain an (a) _____ functional group. Prostaglandins belong to the class of (b) _____ lipids. Simple lipids contain two components, a fatty acid and an (c) _____. Saponifiable lipids may be classified as either simple or (d) _____.

Section 17.2 Fatty Acids

In aqueous solution fatty acid ions form spherical clusters called (a) _____. Fatty acids usually have an (b) _____ number of carbon atoms. Long-chain (c) _____ fatty acids are usually liquids at room temperature. Fatty acids containing no carbon-carbon double bonds are referred to as (d) _____.

Section 17.3 The Structure of Fats and Oils

The alcohol component of a fat is (a) _____. Fats and oils may also be referred to as triacylglycerols or (b) _____. Fats are usually derived from (c) _____ sources. Oils consist of triesters containing (d) _____ fatty acids.

Section 17.4 Chemical Properties of Fats and Oils

A reaction of a fat with water under acidic conditions is called (a) _____. Saponification of an oil produces soaps and (b) _____. Semisolid cooking shortenings are produced from oils by a (c) _____ reaction. Oils containing two or more double bonds are referred to as (d) _____.

Section 17.5 Waxes

The (a) _____ functional group is present in waxes. Waxes consist of an alcohol component and a (b) _____ component. In terms of water solubility, waxes are (c) _____.

233

Section 17.6 Phosphoglycerides

Phosphoglycerides serve as major components of cell (a) _____. The amino alcohol portion of lecithin is (b) _____. Soybean lecithin is used in foods as an (c) _____ agent. Cephalins are particularly abundant in (d) _____ tissue.

Section 17.7 Sphingolipids

Sphingolipids contain an alcohol component called (a) _____. Sphingolipids are abundant in nerve and (b) _____ tissue. A type of sphingolipid which contains a carbohydrate component is called a (c) _____.

Section 17.8 Biological Membranes

Two major cell types are found in living organisms, procaryotic and the more complex (a) _____. Membrane-enclosed bodies within cells are called (b) _____. A widely accepted structure for membranes is called the (c) _____ model. Lipids in a membrane are organized in a (d) _____, a structure consisting of two sheets of lipid molecules arranged so that the hydrophobic portions are facing each other.

Section 17.9 Steroids

The number of rings contained in the steroid ring system is (a) _____. The steroid (b) _____ has been implicated in hardening of the arteries. Steroids emptied from the gallbladder to aid in digestion are called (c) _____. The major component of gallstones is (d) _____.

Section 17.10 Steroid Hormones

A class of adrenocorticoid hormones which regulates the concentration of ions in body fluids is the (a) _____. The hormone which regulates body levels of Na$^+$ is (b) _____. The major glucocorticoid, (c) _____, functions to increase glucose and glycogen concentrations in the body. Female sex hormones are produced in the (d) _____.

Section 17.11 Prostaglandins

Prostaglandins are synthesized from (a) _____ fatty acids. Prostaglandins are similar to (b) _____ in the sense that they are involved in a host of body processes.

SOLUTIONS TO EXERCISES ANSWERED IN THE TEXT

17.1 Lipids are biological molecules soluble in nonpolar solvents.

17.3 ester

17.4 b) saponifiable f) nonsaponifiable
 d) saponifiable

17.6 A micelle is a spherical cluster with the nonpolar portions of fatty acids on the interior and the polar portions on the outside. Dispersion forces hold the nonpolar portions together.

17.8 Unsaturated fatty acids have a bent structure that prevents the molecules from packing as tightly as saturated fatty acids.

17.9 b) unsaturated, liquid d) unsaturated, liquid

17.11
$$\begin{array}{l} CH_2-O-\overset{O}{\underset{\|}{C}}-(CH_2)_{14}CH_3 \\ \;\;\;\;|\;\;\uparrow \\ CH-O-\overset{O}{\underset{\|}{C}}-(CH_2)_7CH=CH(CH_2)_7CH_3 \\ \;\;\;\;|\;\;\uparrow \\ CH_2-O-\overset{O}{\underset{\|}{C}}-(CH_2)_{16}CH_3 \\ \;\;\;\;\;\;\uparrow \end{array}$$

17.13 Animal fat; vegetable oil

17.15 They differ in the nature of the carboxylic acid product. Acid hydrolysis gives a carboxylic acid, and base hydrolysis produces a salt of the carboxylic acid.

17.16 b)

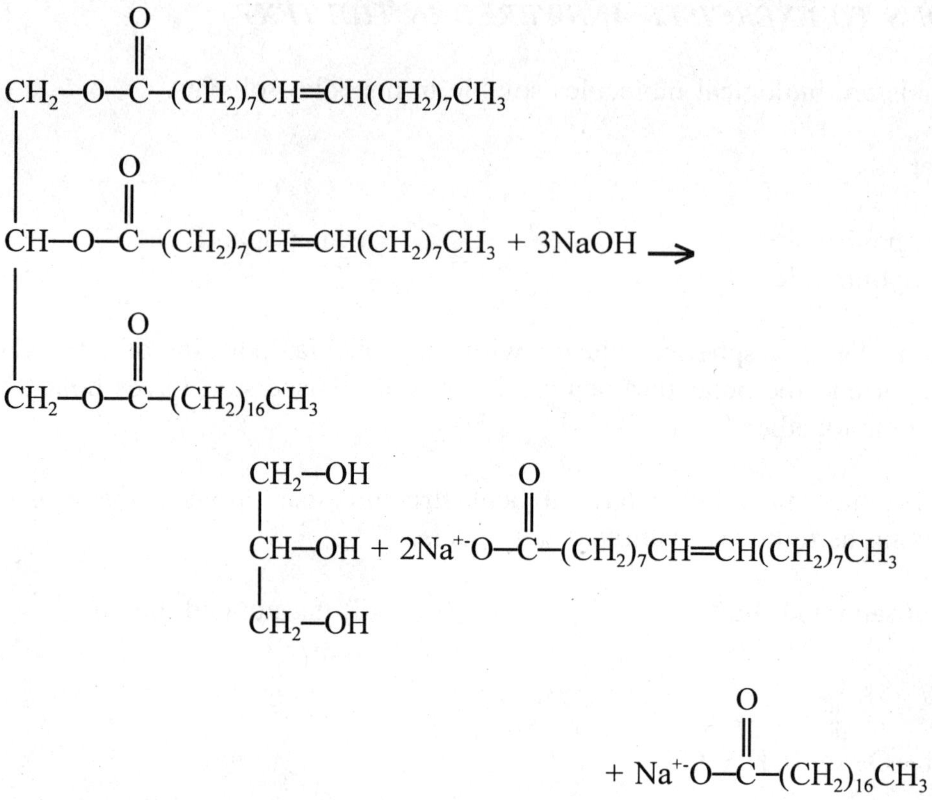

$$\text{CH}_2\text{—O—}\underset{\underset{\text{O}}{\|}}{\text{C}}\text{—(CH}_2)_7\text{CH=CH(CH}_2)_7\text{CH}_3$$
$$|$$
$$\text{CH—O—}\underset{\underset{\text{O}}{\|}}{\text{C}}\text{—(CH}_2)_7\text{CH=CH(CH}_2)_7\text{CH}_3 + 3\text{NaOH} \longrightarrow$$
$$|$$
$$\text{CH}_2\text{—O—}\underset{\underset{\text{O}}{\|}}{\text{C}}\text{—(CH}_2)_{16}\text{CH}_3$$

$$\text{CH}_2\text{—OH}$$
$$|$$
$$\text{CH—OH} + 2\text{Na}^+\text{ }^-\text{O—}\underset{\underset{\text{O}}{\|}}{\text{C}}\text{—(CH}_2)_7\text{CH=CH(CH}_2)_7\text{CH}_3$$
$$|$$
$$\text{CH}_2\text{—OH}$$

$$+ \text{Na}^+\text{ }^-\text{O—}\underset{\underset{\text{O}}{\|}}{\text{C}}\text{—(CH}_2)_{16}\text{CH}_3$$

17.18 $\text{CH}_3(\text{CH}_2)_{14}\text{—}\underset{\underset{\text{O}}{\|}}{\text{C}}\text{—O}^-\text{Na}^+$

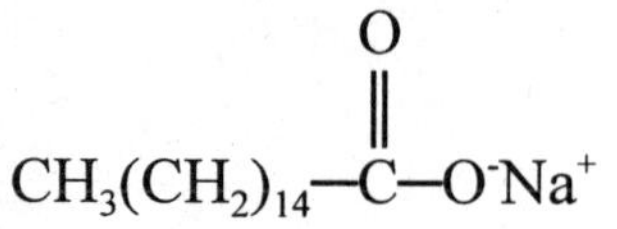

17.19 b)

$$\text{CH}_2\text{—O—}\underset{\underset{\text{O}}{\|}}{\text{C}}\text{—(CH}_2)_{16}\text{CH}_3$$
$$|$$
$$\text{CH—O—}\underset{\underset{\text{O}}{\|}}{\text{C}}\text{—(CH}_2)_{16}\text{CH}_3 + 3\text{H}_2\text{O} \xrightarrow{\text{H}^+}$$
$$|$$
$$\text{CH}_2\text{—O—}\underset{\underset{\text{O}}{\|}}{\text{C}}\text{—(CH}_2)_{16}\text{CH}_3$$

$$\text{CH}_2\text{—OH}$$
$$|$$
$$\text{CH—OH} + 3\text{HO—}\underset{\underset{\text{O}}{\|}}{\text{C}}\text{—(CH}_2)_{16}\text{CH}_3$$
$$|$$
$$\text{CH}_2\text{—OH}$$

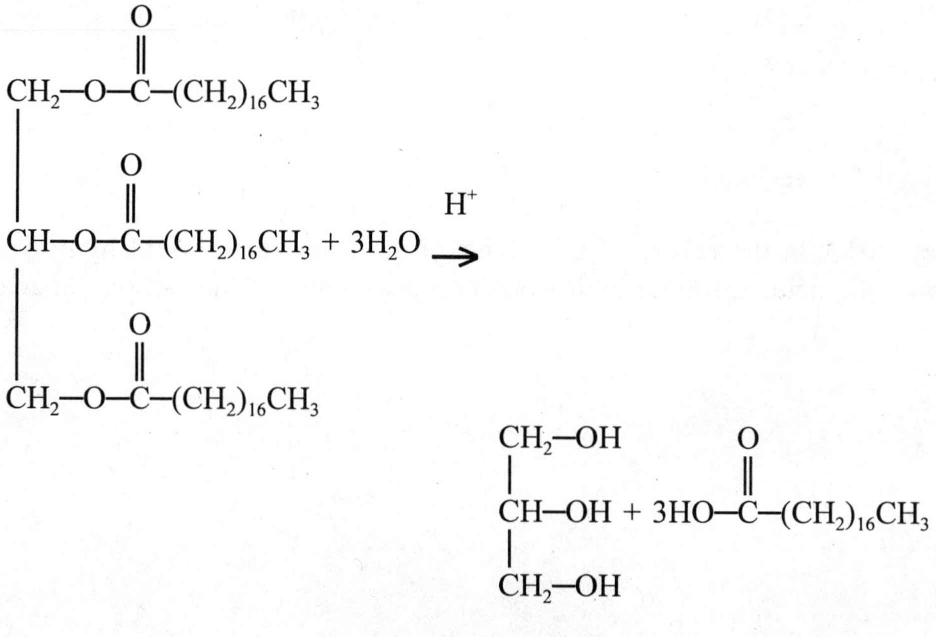

d)
$$CH_2-O-\overset{O}{\underset{\|}{C}}-(CH_2)_{16}CH_3$$
$$CH-O-\overset{O}{\underset{\|}{C}}-(CH_2)_{14}CH_3 + 3NaOH \rightarrow$$
$$CH_2-O-\overset{O}{\underset{\|}{C}}-(CH_2)_{14}CH_3$$

$$\begin{array}{c}CH_2-OH\\|\\CH-OH\\|\\CH_2-OH\end{array} + 2Na^+\,^-O-\overset{O}{\underset{\|}{C}}-(CH_2)_{16}CH_3 + Na^+\,^-O-\overset{O}{\underset{\|}{C}}-(CH_2)_{14}CH_3$$

17.21 $CH_3(CH_2)_{16}-\overset{O}{\underset{\|}{C}}-O-(CH_2)_{15}CH_3$

17.23 Phosphoglycerides contain two components not present in triglycerides: phosphoric acid and an amino alcohol.

17.25 Lecithins serve as components in cell membranes and aid in lipid transport.

17.27 They differ in the identity of the amino alcohol portion. Lecithins contain choline, and cephalins contain ethanolamine or serine.

17.29 Sphingolipids contain the alcohol component sphingosine rather than glycerol, which is present in phosphoglycerides. Sphingolipids contain one fatty acid, and phosphoglycerides contain two fatty acids.

17.31 cerebrosides, brain tissue

17.33 phosphoglycerides, sphingomyelin, and cholesterol

17.35 The model is composed of a flexible bilayer structure with protein molecules floating in the bilayer. The lipid molecules are free to move in a lateral direction.

17.37 bile salts, sex hormones, adrenocorticoid hormones

238 CHAPTER 17

17.39   Bile salts emulsify lipids, providing a much greater surface area for the hydrolysis reactions of digestion.

17.41   The two groups are glucocorticoids and mineralocorticoids. Cortisol, a glucocorticoid, acts to increase the glucose and glycogen concentrations in the body. Aldosterone, a mineralocorticoid, regulates the retention of Na$^+$ and Cl$^-$ during urine formation.

17.43   testosterone; estradial, estrone, and progesterone

17.45   They are cyclopentane derivatives with structures formed from a 20-carbon unsaturated fatty acid.

17.47   They act to regulate menstruation, prevent conception, induce uterine contractions during childbirth, stimulate blood clotting, and control inflammation and fever.

17.49   with premature infants

17.51   Intense pain on the right side, yellow coloration of the skin, or gray-colored stools

17.53   Acne, possible tumors, and infertility

17.55   The end-most double bond is three carbons from the methyl end of the fatty acid.

SELF-TEST QUESTIONS

Multiple Choice

1.   All simple lipids are
     a) salts of fatty acids
     b) esters of fatty acids with various alcohols
     c) esters of fatty acids with alcohol and other additional compounds
     d) esters of fatty acids with glycerol

2.   The esters of fatty acids and alcohols (other than glycerol) are known as
     a) waxes   c) compound lipids
     b) phospholipids   d) fats

3. Which of the following is a glycerol-containing lipid?
   a) sphingolipid
   b) glycolipid
   c) phospholipid
   d) prostaglandin

4. Which of the following is a complex lipid?
   a) steroid
   b) sphingomyelin
   c) prostaglandin
   d) triacylglycerol

5. Which fatty acid is most likely to be found in an oil?
   a) $CH_3(CH_2)_7CH=CH(CH_2)_7COOH$
   b) $CH_3(CH_2)_{14}COOH$
   c) $CH_3(CH_2)_{16}COOH$
   d) $CH_3(CH_2)_{18}COOH$

6. Which of the following food sources would most likely be highest in saturated fatty acids?
   a) cottonseed
   b) corn
   c) beef
   d) sunflower seeds

7. Generally, the structural difference between a fat and an oil is the
   a) alcohol
   b) chain length of fatty acids
   c) degree of fatty-acid unsaturation
   d) degree of fatty-acid chain-branching

8. In triglycerides, fatty acids are joined to glycerol by
   a) ester linkages
   b) ether linkages
   c) phosphate linkages
   d) hydrogen bonds

9. Which of the following is an essential fatty acid?
   a) stearic acid
   b) myristic acid
   c) linoleic acid
   d) palmitic acid

Matching

Match the formulas below to the correct lipid classification given as a response.

10.

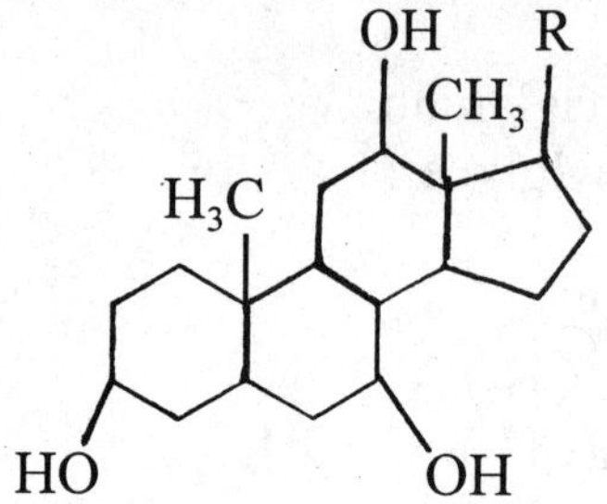

a) steroid
b) complex lipid
c) fat or oil
d) wax

11.

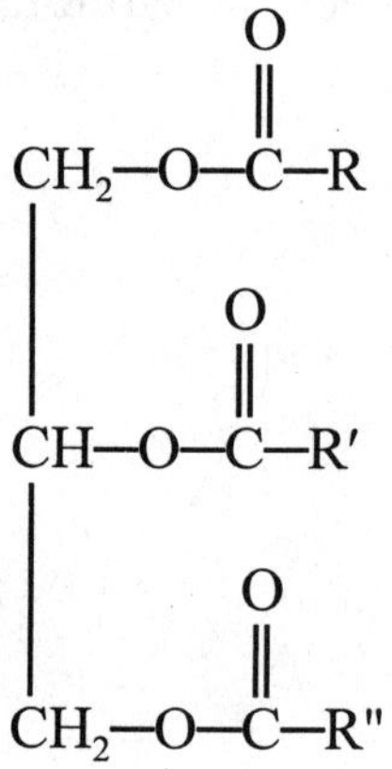

$$\begin{array}{c} CH_2-O-\overset{\overset{O}{\|}}{C}-R \\ | \\ CH-O-\overset{\overset{O}{\|}}{C}-R' \\ | \\ CH_2-O-\overset{\overset{O}{\|}}{C}-R'' \end{array}$$

12. $CH_3(CH_2)_6\overset{\overset{O}{\|}}{C}-O-CH_2(CH_2)_8CH_3$

13.

$$\begin{array}{c} CH_2-O-\overset{\overset{O}{\|}}{C}-R \\ | \\ CH-O-\overset{\overset{O}{\|}}{C}-R' \\ | \\ CH_2-O-\underset{\underset{O^-}{|}}{\overset{\overset{O}{\|}}{P}}-O-R'' \end{array}$$

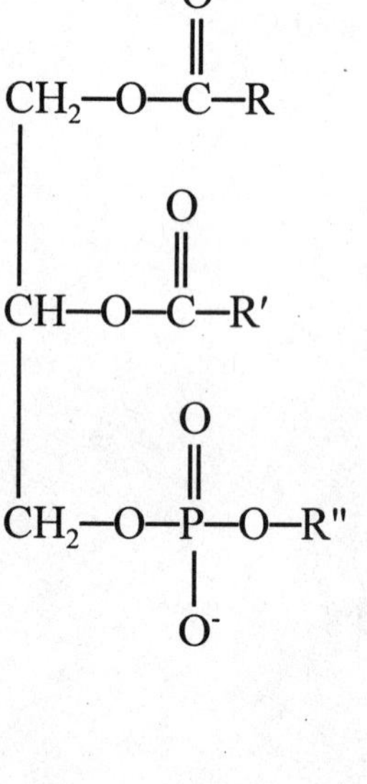

Materials can be obtained from lipids, and the lipids changed by chemical processes. Choose the correct process to accomplish each change described below.

14. obtain a high molecular weight alcohol from a plant wax

15. obtain glycerol from an oil

16. obtain soaps from an oil

17. raise the melting point of an oil

a) hydrogenation
b) acid-catalyzed hydrolysis
c) saponification
d) more than one listed process would work

In the following reaction, L represents a lipid, and A, B, C and D represent possible hydrolysis products.

18. A = fatty acids, B = glycerol, C = phosphoric acid, D = choline

19. A = fatty acids, B = glycerol, no other products

20. A = fatty acids, B = sphingosine, C = phosphoric acid, D = choline

21. A = fatty acids, B = sphingosine, C = carbohydrate, no other product forms

a) simple lipid
b) glycolipid
c) phospholipid
d) sphingolipid

True-False

22. Cell membranes contain about 60% lipid and 40% carbohydrate.

23. Cell membranes are thought to be relatively flexible.

24. A compound containing nine carbon atoms could not be a steroid.

25. In their physiological action, the prostaglandins resemble hormones.

26. All of the male and female sex hormones are steroids.

27. The hormone aldosterone exerts its influence at the pancreas.

ANSWERS TO PROGRAMMED REVIEW

17.1 a) ester   b) nonsaponifiable   c) alcohol   d) complex

17.2 a) micelles   b) even   c) unsaturated   d) saturated

17.3 a) glycerol   b) triglycerides   c) animal   d) unsaturated

17.4 a) hydrolysis   b) glycerol   c) hydrogenation
      d) polyunsaturated

17.5 a) ester   b) fatty acid   c) insoluble

17.6 a) membranes   b) choline   c) emulsifying   d) brain

17.7 a) sphingosine b) brain   c) glycolipid

17.8 a) eucaryotic   b) organelles   c) fluid-mosaic   d) bilayer

17.9 a) four   b) cholesterol   c) bile salts   d) cholesterol

17.10 a) mineralocorticoids   b) aldosterone   c) cortisol
      d) ovaries

17.11 a) unsaturated b) hormones

ANSWERS TO SELF-TEST QUESTIONS

1.	b	10.	a	19.	a
2.	a	11.	c	20.	d
3.	c	12.	d	21.	b
4.	b	13.	b	22.	F
5.	a	14.	d	23.	T
6.	c	15.	d	24.	T
7.	c	16.	c	25.	T
8.	a	17.	a	26.	T
9.	c	18.	c	27.	F

CHAPTER 18

Proteins

PROGRAMMED REVIEW

Section 18.1 The Amino Acids

All twenty common amino acids contain two functional groups, an amino group and a (a) _____ group. Amino acids found in living systems usually exist in the (b) _____ enantiomeric form. Amino acids may be represented by (c) _____-letter abbreviations. There are (d) _____ amino acids which contain an acidic R group.

Section 18.2 Zwitterions

The net charge on a zwitterion is (a) _____. In basic solutions an amino acid has a net (b) _____ charge. The pH at which the zwitterionic form of an amino acid predominates is called the (c) _____.

Section 18.3 Reactions of Amino Acids

The SH-containing amino acid which can be oxidized to a disulfide is (a) _____. The amide linkage between two amino acid components is also called a (b) _____ bond. The amino acid shown on the right side of a peptide is the (c) _____ residue. Polypeptide chains with more than (d) _____ amino acids are usually called proteins.

Section 18.4 Important Peptides

The antidiuretic hormone (ADH) is also known as (a) _____. The pituitary gland synthesizes the (b) _____ hormone which regulates the productivity of the adrenal gland.

Section 18.5 Characteristics of Proteins

Proteins which serve a catalytic function are called (a) _____. Proteins made up of long stringlike molecules are classified as (b) _____ proteins. Proteins comprised solely of amino acids are called (c) _____ proteins. (d) _____ groups are the nonamino acid parts of conjugated proteins.

Section 18.6 The Primary Structure of Proteins

The primary structure of a protein is held together by (a) _____ bonds. The order in which amino acid residues are linked together is the (b) _____ structure of a protein.

Section 18.7 The Secondary Structure of Proteins

The two types of protein secondary structure are the (a) _____ and the (b) _____. Secondary structure of a protein is held intact by (c) _____ bonds.

Section 18.8 The Tertiary Structure of Proteins

The tertiary structure of proteins results from interactions between the
(a) _____ of the amino acids. Interactions between two nonpolar groups are termed
(b) _____ interactions. The attraction of oppositely charged side chains gives rise to
(c) _____ bridges. Two alcohol-containing side chains may interact to form
(d) _____ bonds.

Section 18.9 The Quaternary Structure of Proteins

The polypeptide chains comprising a quaternary structure are called (a) _____. The quaternary structure of hemoglobin consists of (b) _____ chains.

Section 18.10 Protein Hydrolysis and Denaturation

Protein hydrolysis results in the formation of smaller peptides and (a) _____. The natural three-dimensional structure of a protein is called the (b) _____ state.
(c) _____ is the process by which a protein loses its characteristic structure.

SOLUTIONS TO EXERCISES ANSWERED IN THE TEXT

18.1   a carboxylate group and an amino group

18.3   b) thiol group   f) alkyl group
       d) amino group

Proteins

18.4 b)

$$\overset{H}{\underset{CH_2}{\overset{|}{\underset{|}{H_3\overset{+}{N}-C-COO^-}}}}$$
CH₂ — indole ring (tryptophan side chain circled)

d)

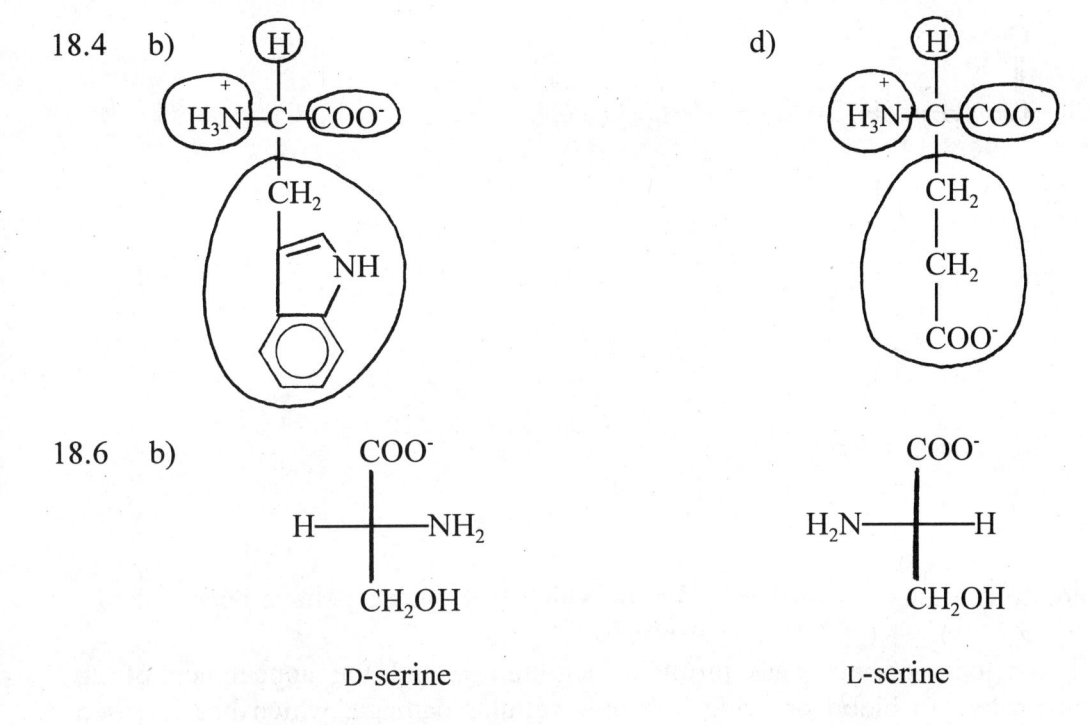

$$H_3\overset{+}{N}-\overset{\overset{H}{|}}{\underset{\underset{\underset{COO^-}{|}}{\underset{CH_2}{|}}}{C}}-COO^-$$

(with CH₂—CH₂—COO⁻ side chain circled)

18.6 b)

```
       COO⁻                    COO⁻
        |                       |
   H ——|—— NH₂            H₂N ——|—— H
        |                       |
       CH₂OH                   CH₂OH

      D-serine                L-serine
```

18.8 crystalline solids with relatively high melting points and high water solubilities

18.10 b) H₂N—CH—COO⁻
              |
             CH—OH
              |
             CH₃

18.11 b) H₃N⁺—CH—COOH
              |
             CH—OH
              |
             CH₃

18.12 b) H₃N⁺—CH—COO⁻ + H₃O⁺ → H₃N⁺—CH—COOH + H₂O
               | |
              CH₂   CH₂
               | |
              CHCH₃   CHCH₃
               | |
              CH₃   CH₃

246 CHAPTER 18

18.14

$$\overset{+}{H_3N}-CH-\overset{O}{\overset{\|}{C}}-NH-CH-\overset{O}{\overset{\|}{C}}-NH-CH-\overset{O}{\overset{\|}{C}}-O^-$$

with side chains: CH₂-C₆H₅ (Phe), CH₂-C₆H₄-OH (Tyr), CH₂-OH (Ser)

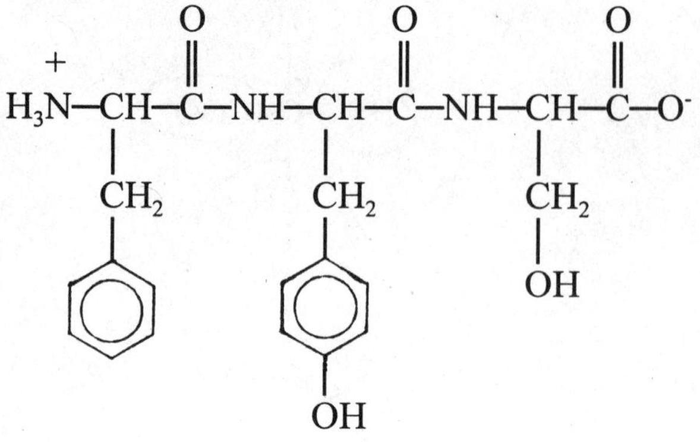

Phe-Tyr-Ser

18.16 only one

18.18 Each peptide contains a disulfide bridge in which two cysteines have been linked.

18.20 Normally proteins do not pass through membranes, and the appearance of an abnormal protein in blood or urine indicates cellular damage, which has released those substances.

18.21

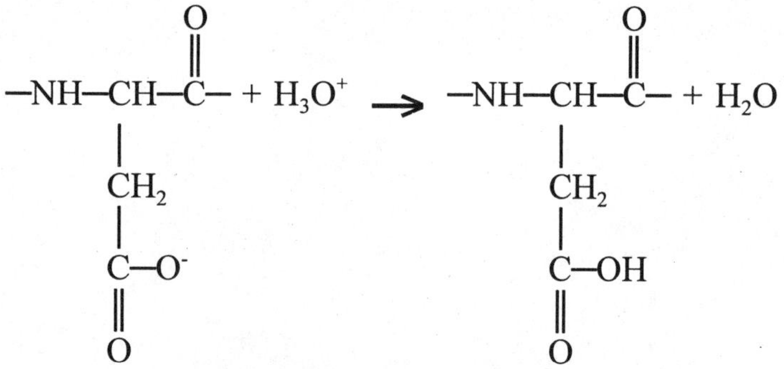

$$-NH-CH-\overset{O}{\overset{\|}{C}}- + H_3O^+ \longrightarrow -NH-CH-\overset{O}{\overset{\|}{C}}- + H_2O$$

with side chains: CH₂-COO⁻ → CH₂-COOH

The aspartate portion of the protein chain would neutralize the H₃O⁺ ions.

18.22 When a protein is at a pH corresponding to the isoelectric point, the net charge on the protein is 0. The protein is not as soluble with a 0 charge as when it possesses a net negative or positive charge.

18.24 catalysis, structural, storage, protection, regulatory, nerve impulse transmission, motion, transport

18.25 b) structural f) transport
      d) catalysis

18.27 b) Ceruloplasmin is a transport protein and is thus dependent on a globular structure.

18.29 b) fibrin   d) mucin

18.31 peptide bonds (amide bonds)

18.33 hydrogen bonding between amide groups

18.35 The α structure is characterized by helical regions in a protein chain. The β structure is a sheetlike array of protein chains aligned side by side.

18.37 Any of the basic amino acids (lysine, arginine, or histidine) coupled with either of the acidic amino acids (aspartate or glutamate).

18.39 b) The side chain of aspartate contains a carboxylate group —COO^-, while the side chain of lysine contains a —NH_3^+ group. Attraction between these two ions produces a salt bridge.

d) Both side chains are neutral and nonpolar. Thus, hydrophobic attractions result.

18.41 a quaternary protein structure is the arrangement of subunits that forms a larger protein

18.43 a subunit is a polypeptide chain that is part of a larger protein

18.45 protein denaturation is the process by which a protein loses its characteristic native structure and function

18.47 both processes result in a denaturation of the fish protein

18.49 detergents can disinfect dishes by virtue of their ability to denature the proteins of microorganisms

18.51 the lead ion

18.53 enkephalins; five

18.55 immunoglobulins form antibodies, which promote the destruction of bacteria and viruses

18.57 Malaria, which is prevalent in some parts of the world, is not as lethal in people carrying the sickle cell trait. Thus, the percentage of the population carrying the sickle cell trait has increased as deaths occurred from malaria.

SELF-TEST QUESTIONS

Multiple Choice

1. The main distinguishing feature between various amino acids is
   a) the length of the carbon chain
   b) the number of amino groups
   c) the identity of the side chain
   d) the number of acid groups

2. The amino acid valine is represented below. Which of the lettered carbon atoms is the alpha carbon atom?

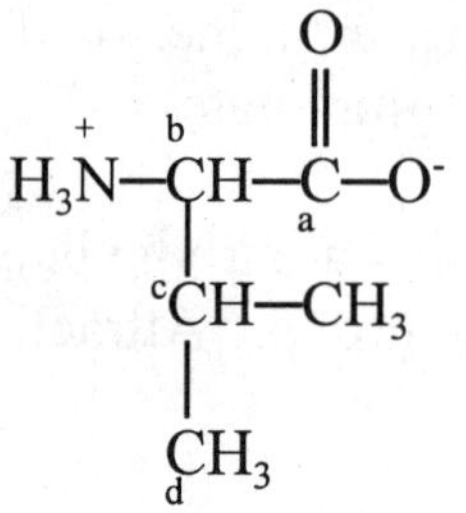

$$\overset{+}{H_3N}-\overset{b}{CH}-\overset{a}{\underset{\underset{\underset{\underset{d}{CH_3}}{|}}{\overset{c}{CH}-CH_3}}{|}}\overset{O}{\overset{\|}{C}}-O^-$$

   a)   b)   c)   d)

3. The compound $\overset{+}{H_3N}-\underset{\underset{\underset{NH_3^+}{|}}{(CH_2)_4}}{\overset{|}{CH}}-\overset{O}{\overset{\|}{C}}-O^-$ is a/an _____ amino acid.

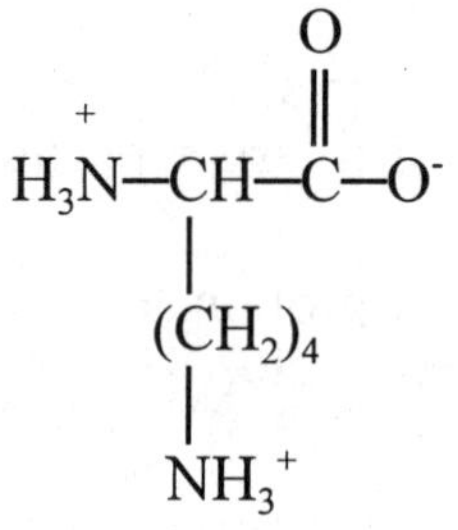

   a) acidic
   b) basic
   c) neutral
   d) more than one response is correct

4. Which of the following would probably represent alanine at its isoelectric point?

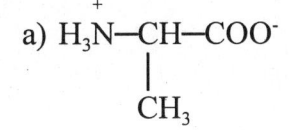

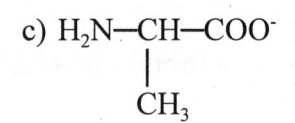

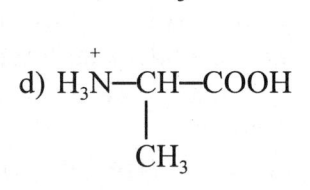

   a) H$_3$N$^+$—CH—COO$^-$
             |
             CH$_3$

   b) H$_2$N—CH—COOH
           |
           CH$_3$

   c) H$_2$N—CH—COO$^-$
           |
           CH$_3$

   d) H$_3$N$^+$—CH—COOH
             |
             CH$_3$

5. A linkage present in all peptides is

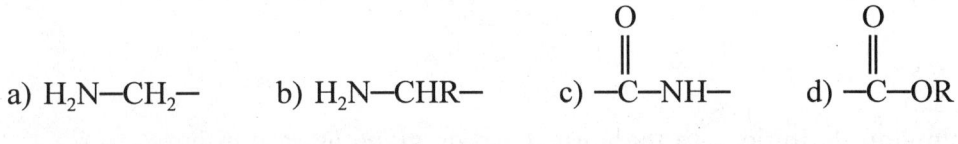

   a) H$_2$N—CH$_2$—   b) H$_2$N—CHR—   c) —C(=O)—NH—   d) —C(=O)—OR

6. Prosthetic groups are found in
   a) all proteins
   b) simple proteins
   c) conjugated proteins
   d) no proteins

7. A protein that is relatively spherical in shape and fairly soluble in water is a _____ protein.
   a) simple
   b) conjugated
   c) fibrous
   d) globular

8. Which of the following characteristics of a protein would be classified as a primary structural feature?
   a) amino acid sequence
   b) pleated-sheet configuration
   c) α-helix configuration
   d) the shape of the protein molecule

9. Which protein serves as antibodies?
   a) myoglobin
   b) hemoglobin
   c) collagen
   d) immunoglobulin

10. Which of the following side-group interactions involves nonpolar groups?
    a) salt bridges
    b) hydrogen bonds
    c) disulfide bonds
    d) hydrophobic bonds

11. Which of the following does *not* contribute to the tertiary structure of proteins?
    a) salt bridges
    b) hydrogen bonds
    c) disulfide bonds
    d) peptide bonds

12. The quaternary structure of hemoglobin involves _____ subunits.
    a) two   b) four   c) six   d) eight

13. Denaturation of a protein involves a breakdown of the
    a) primary structure
    b) secondary and tertiary structures
    c) primary and secondary structures
    d) primary, secondary and tertiary structures

14. Ions of heavy metals (Hg^{2+} or Pb^{2+}) denature proteins by combining with
    a) —NH_2 groups
    b) —SH groups
    c) —OH groups
    d) —C—NH— groups
        $\parallel$
        O

Matching

Match the following definitions to the correct words given as responses.

15. A dipolar amino acid molecule containing both a + and - charge.

16. A substance composed of 25 amino acids linked together.

17. The pH at which amino acids exist in the form that has no net charge.

a) polypeptide
b) ninhydrin
c) isoelectric point
d) zwitterion

Select the correct polypeptide for each description on the left.

18. stimulates milk production

19. controls carbohydrate metabolism

20. decreases urine production

a) myosin
b) insulin
c) oxytocin
d) vasopressin

For each description on the left, select the correct protein class.

21. hemoglobin belongs to this class
22. proteins that function as enzymes
23. insulin belongs to this class
24. keratin belongs to this class
25. collagen belongs to this class

a) regulatory proteins
b) transport proteins
c) structural proteins
d) catalytic proteins

For each of the bonds given on the left, choose a response that indicates the type of protein structure the bond is involved in forming.

26. hydrogen bonds
27. amide bonds
28. hydrophobic bonds
29. salt bonds

a) primary structure
b) secondary structure
c) tertiary structure
d) more than one response is correct

ANSWERS TO PROGRAMMED REVIEW

18.1	a) carboxylate	b) L	c) three	d) two
18.2	a) zero	b) negative	c) isoelectric point	
18.3	a) cysteine	b) peptide	c) C-terminal	d) 50
18.4	a) vasopressin	b) adrenocorticotropic		
18.5	a) enzymes	b) fibrous	c) simple	d) prosthetic
18.6	a) peptide	b) primary		
18.7	a) α-helix	b) β-pleated sheet	c) hydrogen	

18.8   a) R groups   b) hydrophobic   c) salt   d) hydrogen

18.9   a) subunits   b) four

18.10 a) amino acids   b) native   c) denaturation

ANSWERS TO SELF-TEST QUESTIONS

1.	c	11.	d	21.	b
2.	b	12.	b	22.	d
3.	b	13.	b	23.	a
4.	a	14.	b	24.	c
5.	c	15.	d	25.	c
6.	c	16.	a	26.	d
7.	d	17.	c	27.	a
8.	a	18.	c	28.	c
9.	d	19.	b	29.	c
10.	d	20.	d		

CHAPTER 19

Enzymes

PROGRAMMED REVIEW

Section 19.1 The General Characteristics of Enzymes

Enzymes speed up chemical reactions by (a) _____ activation energies. Enzyme (b) _____ is a characteristic of an enzyme that it catalyzes only certain reactions. A third important characteristic of enzymes is that their activity as catalysts can be (c) _____.

Section 19.2 Enzyme Nomenclature and Classification

The (a) _____ is the substance that undergoes a chemical change catalyzed by an enzyme. Systematic names for enzymes end in (b) _____. The IEC classification of enzymes groups them into (c) _____ categories. The common system for naming enzymes incorporates the (d) _____ or the type of reaction into the name.

Section 19.3 Enzyme Cofactors

A (a) _____ is a nonprotein molecule or ion required by an enzyme for catalytic activity. If a molecule required by an enzyme for catalytic activity is organic, it is called a (b) _____. The protein portion of an enzyme which requires an additional molecule or ion for catalytic activity is called the (c) _____.

Section 19.4 Mechanism of Enzyme Action

The (a) _____ is the location on an enzyme where a substrate is bound and catalysis occurs. The combination formed when substrate and enzyme bond is called the enzyme-substrate (b) _____. The (c) _____ theory proposes that a substrate has a shape fitting that of the enzyme's active site. The (d) _____ theory proposes that the conformation of an enzyme changes to accommodate an incoming substrate.

Section 19.5 Enzyme Activity

The number of molecules of substrate acted upon by one molecule of enzyme per minute is the (a) _____ number. Experiments that measure enzyme activity are called enzyme

(b) _____. One standard (c) _____ is the quantity of enzyme which catalyzes the conversion of 1 micromole of substrate per minute.

Section 19.6 Factors Affecting Enzyme Activity

Increasing the concentration of enzyme (a) _____ the rate of an enzyme-catalyzed reaction. As substrate concentration is increased, V_{max} is achieved when the enzyme is (b) _____ with substrate. The temperature at which enzyme activity is highest is the (c) _____ temperature. The (d) _____ pH is that at which enzyme activity is highest.

Section 19.7 Enzyme Inhibition

An (a) _____ is any substance that can decrease the rate of an enzyme-catalyzed reaction. Cyanide ion interferes with the operation of an iron-containing enzyme called (b) _____. An antidote for heavy-metal poisoning is (c) _____. Sulfa drugs are examples of (d) _____ enzyme inhibitors.

Section 19.8 Regulation of Enzyme Activity

A proenzyme or (a) _____ is the inactive precursor of an enzyme. An (b) _____ enzyme is one whose activity is changed by the binding of modulators. Enzyme regulation in which the enzyme that catalyzes the first step of a series of reactions is inhibited by the final product is called (c) _____ inhibition. Enzyme (d) _____ is the synthesis of enzymes in response to a temporary need of the cell.

Section 19.9 Medical Application of Enzymes

An enzyme useful in diagnosing prostate cancer is (a) _____. Multiple forms of the same enzyme are known as (b) _____. Examples of enzymes which occur in multiple forms are CK and (c) _____.

SOLUTIONS TO EXERCISES ANSWERED IN THE TEXT

19.1 Enzymes catalyze the chemical reactions that take place in the body.

19.3 An enzyme lowers the activation energy of the reaction it catalyzes.

19.5 Enzymes are well suited for their role in living organisms because they are powerful (efficient) catalysts, they are highly specific, and their activity as catalysts can be regulated.

19.7 Urea is the substrate acted on by the enzyme urease. Maltose is the substrate acted on by the enzyme maltase.

19.8
Enzyme	Substrate
b) fumarase	a) fumarate
d) arginase	b) arginine

19.9
Enzyme	Reaction
b) phosphatase	a) formation of ester linkages
d) esterase	d) hydrolysis of phosphate ester linkages

19.10 b) cytochrome oxidase: catalyzes the oxidation of cytochrome
d) lactate dehydrogenase: catalyzes the dehydrogenation of lactate

19.11 b) cofactor   d) coenzyme

19.12 An enzyme is a catalytically active structure made up of a protein portion called an apoenzyme and a nonprotein portion called a cofactor.

19.14 Some inorganic ions (minerals) function as cofactors for enzymes.

19.16 E = enzyme, S = substrate, ES = enzyme-substrate complex, and P = product of reaction. Thus, $E + S \rightleftarrows ES \rightarrow E + P$ represents the reaction of an enzyme with a substrate molecule to form an enzyme-substrate complex, which then gives the reaction product and regenerates the enzyme.

19.18 According to the lock-and-key theory, the enzyme surface accommodates only substrates having a specific size and shape. According to the induced-fit theory, the enzyme is somewhat flexible and adapts its shape to accommodate or fit an incoming substrate molecule.

19.20 Any observation that allows the rate of product formation or reactant depletion to be measured will allow enzyme activity to be determined.

19.22 One international unit (IU) is the amount of enzyme that will convert 1 micromol (μmol) of substrate to product in 1 minute. The international unit is useful in medical diagnosis because it represents a standard quantity of enzyme (or enzyme activity level) against which the level present in a patient can be compared.

19.23 b)

[Graph: Rate vs Enzyme concentration — linear increase]

d)

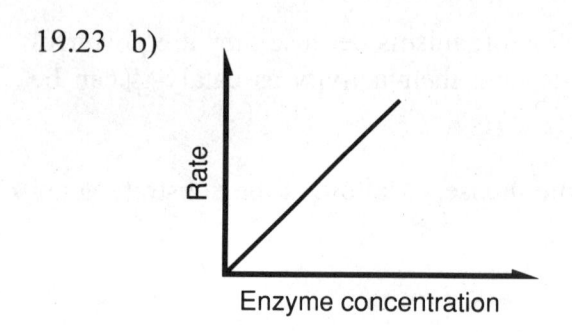

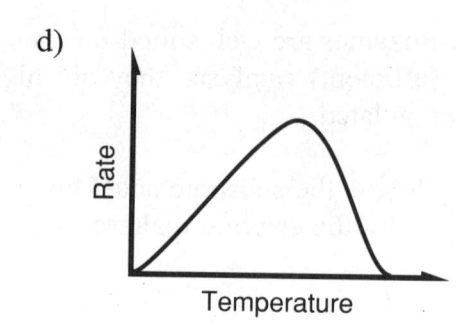

[Graph: Rate vs Temperature — increases to maximum then decreases]

19.24 b) The activity (rate) increases with an increase in enzyme concentration.
d) The activity (rate) increases up to a maximum as the temperature increases and then decreases with further temperature increase.

19.26 Prepare a series of reactions with constant enzyme amount and increasing substrate amounts. Plot reaction rate versus substrate concentration. A constant rate at higher substrate concentrations is V_{max}.

19.28 Enzymes could be destroyed at pH values far from their optimum pH.

19.30 Competitive inhibition results when an inhibitor binds to the active site of an enzyme. Noncompetitive inhibition results when the inhibitor binds reversibly to a site of the enzyme other than the active site. The bonding changes the shape of the enzyme and active site so that the normal substrate no longer fits.

19.31 b) Heavy metal ions combine with the —SH groups found on many enzymes. Some heavy metal ions denature the protein portion of enzymes.

19.32 b) Heavy metal poisoning is treated by administering chelating agents such as EDTA. The chelating agents bond to the metal ions to form water-soluble complexes that are excreted from the body in the urine.

19.33 Enzyme regulation allows an organism to respond to changing conditions by sensitive control of enzyme-catalyzed reactions.

19.35 Enzymes that could attack the cells containing or storing them are often generated and stored in an inactive form called a zymogen. The zymogen is activated only when it is released from storage for use. An example of such an enzyme is the digestive enzyme pepsin.

19.37 allosteric enzymes

Enzymes   257

19.39   In feedback enzyme inhibition, the enzyme that catalyzes the first step in a series of reactions is inhibited by the final product of the series of reactions. As this final product increases in concentration, the inhibition increases, thus preventing the product from being produced in excess of needed amounts.

19.41   Blood serum levels of cellular enzymes increase when the cells are damaged. Thus, enzyme assays of blood serum can help diagnose conditions that cause cell breakdown in specific organs.

19.42   b) GPT assays are useful in diagnosing hepatitis.

19.43   Isoenzymes are different forms of the same enzyme. Two examples of enzymes that occur as isoenzymes are lactate dehydrogenase (LDH) and creatine kinase (CPK).

19.45   The enzymes useful in diagnosing heart attacks are LDH, GOT, and CPK.

19.47   albinism and galactosemia

19.49   Applying lemon juice to sliced fruits and vegetables prevents discoloration by producing acidic conditions that reduce the efficiency of the enzyme polyphenoloxidase.

19.51   The longer the blood supply is impeded, the more widespread is the heart tissue damage.

19.53   Ethanol is acting as a competitive inhibitor through its structural similarities to ethylene glycol.

SELF-TEST QUESTIONS

Multiple Choice

1. Enzymes which act upon only one substance exhibit
   a) catalytic specificity
   b) binding specificity
   c) relative specificity
   d) absolute specificity

2. Which of the following enzyme properties is explained by the lock-and-key model?
   a) specificity
   b) high turnover rate
   c) high molecular weight
   d) susceptibility to denaturation

258 CHAPTER 19

3. The induced-fit theory of enzyme action extends the lock-and-key theory in which of the following ways?
   a) assumes enzymes and substrates are rigid
   b) assumes the shape of substrates changes (conforms) to fit the enzyme
   c) assumes the enzyme shape changes to accommodate the substrate
   d) assumes enzymes have no active site

4. Enzyme turnover numbers are expressed in
   a) activity/mg
   b) units/mg
   c) units/minute
   d) molecules/minute

5. Under saturating conditions, an enzyme-catalyzed reaction had a velocity v. Which of the following would increase the rate of the reaction?
   a) a decrease in the substrate concentration
   b) an increase in the substrate concentration
   c) a decrease in the enzyme concentration
   d) an increase in the enzyme concentration

6. Which of the following has no effect on the rate of an enzyme-catalyzed reaction?
   a) the volume of the reaction mixture
   b) the temperature of the reaction mixture
   c) the pH of the reaction mixture
   d) the enzyme concentration in the reaction mixture

7. In noncompetitive inhibition,
   a) substrate and inhibitor bind at separate locations on the enzyme
   b) substrate and inhibitor bind at the same location on the enzyme
   c) the inhibitor forms a covalent bond with the enzyme
   d) the enzyme becomes permanently inactivated

8. The use of ethanol as a treatment for ethylene glycol poisoning is an excellent example of
   a) feedback inhibition
   b) competitive inhibition
   c) noncompetitive inhibition
   d) enzyme denaturation

9. An enzyme is sometimes generated initially in an inactive form called
   a) a coenzyme
   b) a cofactor
   c) an activator
   d) a zymogen

10. Enzymes which exist in more than one form are called
    a) proenzymes
    b) isoenzymes
    c) apoenzymes
    d) allosteric enzymes

11. Enzymes which play an important role in regulation of enzyme activity are
    a) isoenzymes
    b) proenzymes
    c) allosteric enzymes
    d) apoenzymes

Matching

Select an enzyme that matches each description on the left.

12. name is based on nature of reaction catalyzed
13. name is based on enzyme substrate
14. name is based on both the substrate and the nature of the reaction catalyzed
15. name is an early nonsystematic type

a) pepsin
b) oxidase
c) succinate dehydrogenase
d) sucrase

Match the enzyme components below to the correct term given as a response.

16. the protein portion of an enzyme
17. an organic, but nonprotein portion of an enzyme
18. a nonprotein molecule or ion required by an enzyme

a) cofactor
b) apoenzyme
c) coenzyme
d) proenzyme

Match each description on the left with a response from the right.

19. reacts with cytochrome oxidase
20. stops bacterial synthesis of folic acid
21. inhibits succinate dehydrogenase
22. inhibits bacterial cell-wall construction

a) penicillin
b) cyanide
c) sulfanilamide
d) malonate

Match the correct disease or condition given as a response to the enzyme useful in diagnosing the disease or condition.

23. amylase

24. lysozyme

25. acid phosphatase

a) infectious hepatitis
b) pancreatic diseases
c) prostate cancer
d) monocytic leukemia

True-False

26. The presence of enzymes enables reactions to proceed at lower temperatures.

27. Enzymes are destroyed during chemical reactions in which they participate.

28. Some enzymes are carbohydrates.

29. All enzymes act only on a single substance.

30. Coenzymes are always organic compounds.

31. The optimum pH of all enzymes in the body is near pH 7.

32. A plot of reaction rate (vertical axis) versus temperature of enzyme-catalyzed reactions most frequently yields a straight line.

33. Streptokinase is useful in dissolving blood clots.

34. Lactate dehydrogenase exists in five different isomeric forms.

ANSWERS TO PROGRAMMED REVIEW

19.1 a) lowering   b) specificity   c) regulated

19.2 a) substrate   b) -ase   c) six   d) substrate

19.3 a) cofactor   b) coenzyme   c) apoenzyme

19.4 a) active site   b) complex   c) lock-and-key   d) induced-fit

19.5 a) turnover   b) assays   c) international unit

19.6 a) increases   b) saturated   c) optimum   d) optimum

19.7 a) inhibitor   b) cytochrome oxidase   c) EDTA
      d) competitive

19.8 a) zymogen   b) allosteric   c) feedback   d) induction

19.9 a) acid phosphatase   b) isoenzymes   c) LDH

ANSWERS TO SELF-TEST QUESTIONS

1.	d	13.	d	24.	d
2.	a	14.	c	25.	c
3.	c	15.	a	26.	T
4.	d	16.	b	27.	F
5.	d	17.	c	28.	F
6.	a	18.	a	29.	F
7.	a	19.	b	30.	T
8.	b	20.	c	31.	F
9.	d	21.	d	32.	F
10.	b	22.	a	33.	T
11.	c	23.	b	34.	T
12.	b				

CHAPTER 20

Nucleic Acids and Protein Synthesis

PROGRAMMED REVIEW

Section 20.1 Components of Nucleic Acids

The monomers of nucleic acids are (a) _____. The sugar present in RNA is (b) _____. Adenine is one of two (c) _____ bases. Uracil is one of three (d) _____ bases.

Section 20.2 Structure of DNA

The nucleic acid "backbone" consists of a phosphate- (a) _____ chain. Phosphate groups in DNA link the 3′ position of one sugar to the (b) _____ position of the next sugar. Two DNA strands with matched sequences are said to be (c) _____ to each other. The 3′ end of the sequence AGC is located at the letter (d) _____.

Section 20.3 Replication of DNA

The hereditary material within a human cell consists of 46 packets of DNA called (a) _____. The process by which an exact copy of a DNA molecule is produced is called (b) _____. The point at which a DNA molecule unwinds in the copying process is called a (c) _____. DNA segments produced during the copying process are called (d) _____ fragments.

Section 20.4 Ribonucleic Acid (RNA)

The most abundant type of RNA in a cell is (a) _____. The region of a tRNA molecule that binds to mRNA during protein synthesis is called the (b) _____. The type of RNA which consists of two subunits is (c) _____. (d) _____ functions as a carrier of genetic information from the DNA of the cell nucleus to the site of protein synthesis.

264 CHAPTER 20

Section 20.5 The Flow of Genetic Information

The flow of genetic information according to the (a) _____ dogma of molecular biology is from DNA to (b) _____ to protein. The process by which genetic information is passed from DNA to mRNA is called (c) _____. The process by which a specific protein results from information carried on mRNA is called (d) _____ of the code.

Section 20.6 Transcription: RNA Synthesis

The enzyme (a) _____ catalyzes the synthesis of RNA. Newly formed RNA is synthesized in the (b) _____ to _____ direction. Eucaryotic DNA segments that carry no amino acid code are called (c) _____ while coded segments are called (d) _____.

Section 20.7 The Genetic Code

Code words on mRNA are known as (a) _____. Code words consist of a sequence of (b) _____ nucleotide bases on mRNA. Of the 64 possible code words, 61 represent amino acids. The remaining three are signals for chain (c) _____. Most of the amino acids are represented by more than one codon, a condition known as (d) _____.

Section 20.8 Translation and Protein Synthesis

The first amino acid to be involved in protein synthesis in procaryotic cells is (a) _____. The (b) _____ site is where an incoming tRNA carrying an amino acid attaches to the mRNA-ribosome complex. The movement of a ribosome along a mRNA is called (c) _____. Complexes of several ribosomes and mRNA are called polyribosomes or (d) _____.

Section 20.9 Mutations

A mutation is any change resulting in an incorrect sequence of bases on (a) _____. A mutation can lead to an incorrect sequence of (b) _____ for a protein. Chemicals that induce mutations are called (c) _____.

Section 20.10 Recombinant DNA

The application of recombinant DNA technology is sometimes referred to as (a) _____ engineering. Protective enzymes which can catalyze the cleaving of foreign DNA are called (b) _____ enzymes. A carrier of foreign DNA into a cell is called a (c) _____. Circular DNA often used as a carrier is called a (d) _____.

SOLUTIONS TO EXERCISES ANSWERED IN THE TEXT

20.2 Nucleotides are the repeating structural units or monomers of polymeric nucleic acids.

20.4 b) pyrimidine   d) pyrimidine

20.5 Bases of DNA: thymine, cytosine, adenine, guanine
Bases of RNA: uracil, cytosine, adenine, guanine

20.7 DNA molecules containing the same number of nucleotides could differ in the bases contained in the nucleotides.

20.9 Watson and Crick proposed a double-stranded helical structure for DNA, with the two strands held together by hydrogen bonds between complementary base pairs extending across the axis of the helix.

20.11 Hydrogen bonds between complementary base pairs extend from one DNA strand to the other through the axis of the helix. These bonds hold the strands together in the helical configuration.

20.12 b) Each complementary base pair (AT or TA) involves 2 hydrogen bonds. There would be six base pairs held together by 2 x 6 or 12 hydrogen bonds.

20.13 The original sequence GCTTAG is written in the 5′ → 3′ direction. The complementary sequence is CGAATC but it is antiparallel (written in the 3′ → 5′ direction). Turning the complementary sequence around to 5′ → 3′ direction gives CTAAGC.

20.15 *Step 1*: unwinding the DNA double helix
Step 2: synthesis of DNA segments
Step 3: closing the nicks in one of the new daughter strands

20.17 DNA polymerase and DNA ligase

20.19 Daughter strands growing toward replication forks grow smoothly in a continuous chain as new forks are exposed. Daughter strands growing away from replication forks grow in segments because a new strand segment begins at each new fork. These small segments are later joined to form the complete daughter strand.

20.22 The sugar in the RNA backbone is ribose, and that in the DNA backbone is deoxyribose.

20.24 The ratio of guanine to cytosine in RNA does not have to be 1:1 because parts of the single strand do not fold back to form helices. The 1:1 ratio is necessary only in those parts that do form helices. Thus, the overall ratio can be different than 1:1.

20.26 The two important regions of a tRNA molecule are the anticodon, which binds the tRNA to mRNA during protein synthesis, and the 3′ end, which binds to an amino acid by an ester linkage.

20.28 Transcription is the transferring of genetic DNA information to messenger RNA. Translation is the expression of messenger RNA information in the form of an amino acid sequence in a protein.

20.29 *Step 1*: A DNA double helix unwinds to expose a DNA region that is to be transcribed.
Step 2: Ribonucleotides line up along the exposed DNA strand through complementary base pairing.
Step 3: RNA polymerase catalyzes formation of the RNA strand by linking ribonucleotides.

20.31 Transcription of a DNA molecule in procaryotes produces a mRNA that is immediately translated into a protein. Transcription in eucaryotes produces heterogeneous nuclear RNA, containing introns and exons. The hnRNA is cut and spliced to produce a mRNA molecule that then leaves the nucleus for the site of protein synthesis in the cytoplasm.

20.33 The use of three-base words allows enough words to be formed so that each amino acid can be represented uniquely by at least one word. A code using four or more bases per word would work but would form many more words than does a three-base code. With four bases available, four-letter words would produce a code containing 256 words.

20.35 A synthetic mRNA was made that contained only one base (uracil). When this synthetic mRNA was put into an environment containing the components needed to synthesize a protein, the protein that resulted contained only the amino acid phenylalanine. Thus, the codon UUU corresponded to the amino acid phenylalanine.

20.36 b) true
d) False; each living species is thought to use the same genetic code.
f) False; stop signals for protein synthesis are represented by three codons: UAA, UAG, and UGA.

20.37 The anticodon is the region of a tRNA molecule that binds to the codon of an mRNA molecule.

Nucleic Acids and Protein Synthesis 267

20.39. *N*-terminal amino acid

20.41 DNA segment for transcription
      3'   TAC-CCA-AAT-GGA-ACA   5'
      5'   AUG-GGU-UUA-CCU-UGU   3'
mRNA segment transcribed
Step 1:   The DNA double helix unwinds to expose the gene segment above.
Step 2:   The mRNA sequence shown is synthesized and leaves the nucleus.
Step 3:   A ribosome attaches to the mRNA at the initiation site.
Step 4:   The ribosome moves along the mRNA in the 5' to 3' direction as tRNA molecules bring the proper amino acids to the site of protein synthesis by matching the tRNA anticodons with mRNA codons.
Step 5:   Amino acids are linked by amide bond formation.

20.43 a) A mutation can lead to an incorrect amino acid sequence for a protein. The abnormal protein may no longer be effective.

20.46 Restriction enzymes are used to cleave DNA into gene segments and to cleave plasmids. DNA ligase is used to seal a gene segment into a cleaved plasmid. The result of these processes is that a circular plasmid with a new gene segment is produced.

20.48 Human insulin will become widely available for treating diabetes. Human growth hormone will be used for treating pituitary disorders. Recombivax Hb, a vaccine for hepatitis B, will be marketed.

20.49 Viruses take over the protein-synthesizing machinery of a host cell and cause the cell to produce viral components at the expense of normal cell components.

20.51 A retrovirus contains RNA rather than DNA and uses the RNA to direct the synthesis of DNA within a host cell.

20.53 DNA fingerprinting can link a suspect to evidence such as blood, semen, or saliva found at a crime scene.

20.55 100,000

SELF-TEST QUESTIONS

Multiple Choice

1. A primary difference between DNA and RNA involves
   a) the phosphate
   b) the pentose sugar
   c) the type of bonding between sugar and base
   d) the type of bonding between sugar and phosphate

2. Which of the following bases is a purine?
   a) uracil
   b) adenine
   c) thymine
   d) cytosine

3. A sample of nucleic acid is to be analyzed. For which of the following would an analysis be useful in deciding whether the sample was DNA or an RNA?
   a) guanine
   b) cytosine
   c) adenine
   d) thymine

4. The structural backbone of all nucleic acids consists of alternating molecules of
   a) a sugar and a base
   b) a base and phosphate
   c) a sugar and phosphate
   d) purine and pyrimidine bases

5. If the nucleotide sequence in one strand of DNA were T-C-G, the complementary strand would be
   a) A-C-G   b) T-G-C   c) C-G-A   d) A-G-C

6. A sample of double helical DNA contained 26% of the base thymine (T). The amount of adenine (A) should be
   a) 26%   b) 24%   c) 22%   d) 20%

7. The number of hydrogen bonds between A and T in double-stranded helical DNA is
   a) 1   b) 2   c) 3   d) 4

8. mRNA synthesis is called
   a) replication
   b) translocation
   c) translation
   d) transcription

9. Histidine is represented by the codons CAC and CAU. This is an example of
   a) degeneracy of the genetic code
   b) a nonsense codon
   c) punctuation in the genetic code
   d) the universal nature of the genetic code

10. The genetic code consists of _____ three-letter codons.
    a) 16   b) 32   c) 64   d) 86

11. Segments of eucaryotic DNA which are coded for amino acids are called
    a) exons   b) introns   c) hnRNA   d) codons

12. Recombinant DNA is DNA
    a) formed by combining portions of DNA from two different sources
    b) found within plasmids
    c) with an unusual ability to recombine
    d) which contains two template strands

Matching

Characteristics are given below for various nucleic acids. Match each description to the correct type of nucleic acid as a response.

13. represents a large percent of cellular RNA

14. serves as master template for the formation of all nucleic acids in the body

15. contains base sequences called codons

16. has lowest molecular weight of all nucleic acids

17. replicated during cell division

18. delivers amino acids to site of protein synthesis

19. contains anticodons

20. helps to serve as a site for protein synthesis

21. carries the directions for protein synthesis to the site of protein synthesis

a) DNA
b) mRNA
c) tRNA
d) rRNA

True-False

22. Purine bases contain both a five- and a six-membered ring.

23. DNA and RNA contain identical nucleic acid backbones.

270 CHAPTER 20

24. During DNA replication, only one DNA strand serves as a template for the formation of a complementary strand.

25. A complementary DNA strand formed during replication is identical to the strand serving as a template.

26. RNA is usually double-stranded.

27. Some amino acids are represented by more than one codon.

28. The genetic code has some code words that don't code for any amino acids.

29. The genetic code is thought to be different for humans and bacteria.

30. Incoming tRNA, carrying an amino acid, attach at the peptide site.

31. *N*-formylmethionine is always the first amino acid in a growing peptide chain in procaryotic cells.

32. Protein synthesis generally takes place within the cell nucleus.

33. Some genetic mutations might aid an organism rather than hinder it.

34. Viruses contain both nucleic acids and protein.

35. Restriction enzymes act at sites on DNA called palindromes.

ANSWERS TO PROGRAMMED REVIEW

20.1 a) nucleotides   b) ribose   c) purine   d) pyrimidine

20.2 a) sugar   b) 5′   c) complementary   d) C

20.3 a) chromosomes   b) replication   c) replication fork
      d) Okazaki

20.4 a) rRNA   b) anticodon   c) rRNA   d) mRNA

20.5 a) central   b) mRNA   c) transcription   d) translation

20.6 a) RNA polymerase   b) 5′, 3′   c) introns   d) exons

20.7 a) codons   b) three   c) termination   d) degeneracy

20.8 a) methionine   b) A   c) translocation d) polysomes

20.9 a) DNA   b) amino acids   c) mutagens

20.10 a) genetic   b) restriction   c) vector   d) plasmid

ANSWERS TO SELF-TEST QUESTIONS

1.	b	13.	d	25.	F
2.	b	14.	a	26.	F
3.	d	15.	b	27.	T
4.	c	16.	c	28.	T
5.	c	17.	a	29.	F
6.	a	18.	c	30.	F
7.	b	19.	c	31.	T
8.	d	20.	d	32.	F
9.	a	21.	b	33.	T
10.	c	22.	T	34.	T
11.	a	23.	F	35.	T
12.	a	24.	F		

CHAPTER 21

Nutrition and Energy for Life

PROGRAMMED REVIEW

Section 21.1 Nutritional Requirements

Nutrients required in relatively large amounts by the body are called (a) _____. Those required in only small amounts are called (b) _____. Nutritional guidelines designed to maintain good health for 95% of the U.S. population are called (c) _____ _____ _____ and abbreviated (d) _____.

Section 21.2 The Macronutrients

Nutrients classified as macronutrients include (a) _____, (b) _____ and (c) _____. Dietary carbohydrates are often classified as (d) _____ or (e) _____. The (f) _____ carbohydrates consist primarily of (g) _____, while the (h) _____ carbohydrates are collectively called (i) _____. Most of the lipids in food and in the body are (j) _____. Lipids called oils usually contain a high percentage of (k) _____ fatty acids. Linoleic acid is an example of an (l) _____ _____ _____. Proteins in food that contain all the (m) _____ amino acids in the proportions needed by the body are called (n) _____ proteins.

Section 21.3 Micronutrients I: Vitamins

(a) _____ are organic micronutrients that the body cannot produce in amounts needed for good health. The highly (b) _____ nature of the molecules of water-soluble vitamins renders them water soluble. All water-soluble vitamins except (c) _____ have been shown to function as (d) _____. Fat-soluble vitamins have very (e) _____ molecular structures that cause them to be insoluble in water, but soluble in the body lipids called (f) _____. The four fat-soluble vitamins are (g) _____.

Section 21.4 Micronutrients II: Minerals

(a) _____ are metals or nonmetals used in the body in the form of ions or compounds. (b) _____ minerals are found in the body in quantities greater than 5 g, while

273

274 CHAPTER 21

(c) _____ minerals are found in quantities (d) _____ than 5 g. The inorganic structural components and principal ions of the body utilize (e) _____ minerals, while (f) _____ minerals are important in some enzymes, vitamins and hormones.

Section 21.5 The Flow of Energy in the Biosphere

The (a) _____ is the ultimate source of energy used in all biological processes. The process of photosynthesis requires energy, CO_2 and (b) _____. The products of photosynthesis are carbohydrates and (c) _____. As plants and animals combine carbohydrates with oxygen the products are (d) _____ and water.

Section 21.6 Metabolism and an Overview of Energy Production

The sum total of all the chemical reactions involved in maintaining a living cell is called (a) _____. (b) _____ consists of all reactions that lead to the breakdown of biomolecules. (c) _____ includes all reactions that lead up to the synthesis of biomolecules. Stage III in the extraction of energy from food is referred to as the common (d) _____ pathway. The whole purpose in the extraction of energy from food is to convert the chemical energy in foods to molecules of (e) _____.

Section 21.7 ATP: The Primary Energy Carrier

The base in ATP is (a) _____. Phosphate (P_i) is sometimes referred to as (b) _____ phosphate. If a hydrolysis reaction is energy-releasing, ΔG has a (c) _____ sign. Compounds that liberate much free energy upon hydrolysis are called (d) _____ compounds.

Section 21.8 Coenzymes Important in the Common Catabolic Pathway

Stage II of the oxidation of foods produces (a) _____. Coenzyme A is derived from the B vitamin (b) _____. NAD^+ is derived from the vitamin (c) _____. The other major electron carrier (in addition to NAD^+) in the oxidation of fuel molecules is (d) _____.

SOLUTIONS TO EXERCISES ANSWERED IN THE TEXT

21.1   The primary difference between a macronutrient and a micronutrient is the amount required by the body. Macronutrients are required in relatively large amounts, whereas only small amounts of micronutrients are needed.

21.3   fiber provides bulk for the feces

21.5   b) any of the following: grains (wheat, rice, corn, oats, barley), potatoes, yams, or sweet potatoes
       d) any of the following: lean meat, beans, eggs, fish, and poultry

21.6   b) carbohydrates, lipids, proteins   f) lipids, proteins, carbohydrates
       d) carbohydrates, lipids, proteins

21.9   they serve as coenzymes

21.11 b) vitamin K   d) vitamin E

21.12 b) vitamin B_1 d) niacin

21.14 minerals are classified as major or trace on the basis of the amount present in the body

21.16 the general functions of trace minerals in the body are to serve as components of vitamins (Co), enzymes (Zn, Sc), hormones (I), or specialized proteins (Fe, Cu)

21.18 $6CO_2 + 6H_2O \rightarrow C_6H_{12}O_6 + 6 O_2$

21.20 Energy from the sun powers the process of photosynthesis in plants. As the carbohydrates of plants are converted to CO_2 and H_2O, the released energy is trapped in molecules of ATP.

21.22 Stage I is the digestion of food to provide fatty acids, glycerol, glucose, and amino acids. Stage II is the conversion of these substances to acetyl CoA. Stage III consists of the citric acid cycle followed by electron transport and oxidative phosphorylation where the energy released is trapped in molecules of ATP.

21.23 b) I   d) II

21.24 the reactions of stage III are the same regardless of the type of food being degraded

21.26 The energy given off during catabolism of fuel molecules is saved by molecules of ATP. The ATP molecules act as energy carriers and deliver the energy to where it is needed.

21.28 P_i is the symbol for phosphate ion. PP_i denotes the pyrophosphate ion.

276   CHAPTER 21

21.30 $\Delta G^{\circ\prime}$ is the change in free energy in a reaction at cellular conditions of pH, temperature, and concentration. If $\Delta G^{\circ\prime}$ is negative the reaction is exergonic; if $\Delta G^{\circ\prime}$ is positive, it is endergonic.

21.32   the triphosphate portion

21.34   The ATP-ADP cycle is the link between energy supplied to cells and the energy used by cells. Oxidation of fuel molecules provides energy for the conversion of ADP to ATP. Then ATP provides energy for cellular processes as it returns to ADP.

21.36   Mitochondria are cigar-shaped organelles with both an outer and inner membrane. The inner membrane has many folds and contains the enzymes for the electron transport chain and oxidative phosphorylation. The enzymes for the citric acid cycle are attached or near to the surface of the inner membrane.

21.37   b) FAD serves as a transporter of hydrogen atoms and electrons from the oxidation of fuel molecules to the electron transport chain

21.38   b) riboflavin

21.41   b) $CH_3-OH + NAD^+ \rightarrow H-\overset{\overset{\displaystyle O}{\|}}{C}-H + NADH + H^+$

21.43   it would decrease the caloric content of those foods

21.45   no, exercise builds muscle

21.47   biosynthesis, mechanical, and chemical transport

SELF-TEST QUESTIONS

Multiple Choice

1.   Nutrients required by the body in relatively large amounts are
     a) fiber   c) micronutrients
     b) macronutrients   d) vitamins

2. The "D" in RDA stands for
   a) daily
   b) determined
   c) dietary
   d) deficiency

3. Which of the following is an example of a complex carbohydrate nutrient?
   a) starch
   b) cellulose
   c) glucose
   d) lactose

4. Which of the following has an established RDA?
   a) carbohydrates
   b) lipids
   c) proteins
   d) fiber

5. Which of the following is a water-soluble vitamin?
   a) vitamin E
   b) vitamin B_2
   c) vitamin D
   d) vitamin A

6. Which of the following is a trace mineral?
   a) phosphorus
   b) sulfur
   c) potassium
   d) copper

7. It is recommended that no more than ____ percent of our calories be obtained from fats.
   a) 10
   b) 30
   c) 40
   d) 50

8. Mitochondria are called "power stations of the cell" because
   a) most of the cellular energy is consumed there
   b) mitochondria are rich in fat molecules
   c) mitochondria exist at higher temperatures than other cell components
   d) mitochondria are the major sites of ATP synthesis

9. What product of the second stage is passed on to the third stage of the energy production processes?
   a) CO_2
   b) H atoms
   c) acetyl CoA
   d) H_2O

10. FAD often accompanies the enzyme-catalyzed
    a) oxidation of —CH_2CH_2— to —CH=CH—
    b) oxidation of alcohol groups
    c) oxidation of aldehydes
    d) transfer of acetyl groups

Matching

Select the stage in the extraction of energy from food where the following molecules are produced.

11. majority of ATP

12. amino acids

13. acetyl CoA

14. carbon dioxide

a) stage I
b) stage II
c) stage III
d) more than one response is correct

Select the mineral that best matches each description on the left.

15. a major mineral found in the body in largest amounts

16. a vitamin component

17. a hormone component

a) cobalt
b) iodine
c) calcium
d) zinc

Select the coenzyme that best matches each description on the left.

18. serves as an electron acceptor in the oxidation of an alcohol

19. a derivative of the vitamin riboflavin

20. contains a sulfhydryl group (—SH)

a) coenzyme Q
b) coenzyme A
c) FAD
d) NAD^+

Select the process that best matches each description on the left.

21. energy-releasing solar process

22. conversion of CO_2 and H_2O to carbohydrates

23. oxidation of glucose to CO_2 and H_2O

a) cellular respiration
b) photosynthesis
c) nuclear fusion
d) cellular anabolism

True-False

24. Carbon dioxide is a source of carbon for the earth's organic compounds.

25. ΔG for the hydrolysis of ATP is a positive value.

26. Ten amino acids are listed as being essential.

27. The base present in ATP is deoxyribose.

28. An ATP molecule at pH 7.4 (body pH) has a -3 charge.

29. On a mass basis, water is the most abundant compound found in the human body.

30. Sucrose, a disaccharide, is correctly classified as a complex carbohydrate.

31. Most nutritional studies indicate that typical diets in the U.S. contain too high a percentage of complex carbohydrates.

32. Lipids are digested faster than carbohydrates or proteins.

33. Concern about vitamin overdoses focuses primarily on water-soluble vitamins.

34. Most fat-soluble vitamins are known to function in the body as coenzymes.

35. Vitamin K is important in the process of blood clotting.

ANSWERS TO PROGRAMMED REVIEW

21.1 a) macronutrients   b) micronutrients   c) recommended dietary allowances   d) RDA

21.2 a) carbohydrates   b) lipids   c) proteins   d) simple
      e) complex   f) simple   g) sugars   h) complex
      i) starch   j) triglycerides   k) unsaturated
      l) essential fatty acid   m) essential   n) complete

21.3 a) vitamins   b) polar   c) vitamin C   d) coenzymes
      e) nonpolar   f) fats   g) ADEK

21.4 a) minerals   b) major   c) trace   d) less   e) major   f) trace

21.5 a) sun   b) H_2O c) O_2 d) CO_2

21.6 a) metabolism   b) catabolism   c) anabolism   d) catabolic   e) ATP

21.7 a) adenosine   b) inorganic   c) negative   d) high-energy

21.8 a) acetyl coenzyme A   b) pantothenic acid   c) nicotinamide   d) FAD

ANSWERS TO SELF-TEST QUESTIONS

1.	b	13.	b	25.	F
2.	c	14.	c	26.	T
3.	a	15.	c	27.	F
4.	c	16.	a	28.	F
5.	b	17.	b	29.	T
6.	d	18.	d	30.	F
7.	b	19.	c	31.	F
8.	d	20.	b	32.	F
9.	c	21.	c	33.	F
10.	a	22.	b	34.	F
11.	c	23.	a	35.	T
12.	a	24.	T		

CHAPTER 22

Carbohydrate Metabolism

PROGRAMMED REVIEW

Section 22.1 Digestion of Carbohydrates

The major function of dietary carbohydrate is to serve as an (a) _____ source. Digestion of polysaccharides produces (b) _____. The focal point of carbohydrate metabolism is the monosaccharide (c) _____.

Section 22.2 Blood Glucose

The amount of glucose in the blood is referred to as the blood sugar (a) _____. If a blood sugar concentration is below normal, a condition called (b) _____ exists. When the blood glucose concentration is above normal, the condition is referred to as (c) _____. The blood glucose concentration at which glucose is excreted in the urine is called the (d) _____.

Section 22.3 Glycolysis

Glycolysis is the conversion of glucose to (a) _____. Glycolysis occurs within the (b) _____ of the cell. The glycolysis pathway is regulated by three (c) _____. The phosphorylation of glucose is controlled by (d) _____ inhibition.

Section 22.4 Fates of Pyruvate

Under aerobic conditions, pyruvate is converted to (a) _____. Under anaerobic conditions within the body, pyruvate is converted to (b) _____. Alcoholic fermentation involves the conversion of glucose to (c) _____. Each fate of pyruvate involves the regeneration of (d) _____ so that glycolysis can continue.

Section 22.5 The Citric Acid Cycle

The fuel of the citric acid cycle is (a) _____. Carbon atoms leave the cycle as molecules of (b) _____. Four oxidation-reduction reactions in the cycle produce (c) _____ molecule(s) of NADH and (d) _____ molecule(s) of $FADH_2$.

Section 22.6 The Electron Transport Chain

The enzymes for the electron transport chain are located within the inner membrane of the (a) _____. Molecular oxygen is reduced in the electron transport chain and (b) _____ is formed. A group of iron-containing enzymes called (c) _____ are located in the electron transport chain. Electrons are brought to the electron transport chain by $FADH_2$ and (d) _____.

Section 22.7 Oxidation Phosphorylation

Oxidative phosphorylation occurs at (a) _____ different locations of the electron transport chain. During oxidative phosphorylation ADP is converted to (b) _____. The theory which proposes a mechanism for oxidative phosphorylation is called the (c) _____ hypothesis. This hypothesis proposes that a (d) _____ flow occurs across the inner mitochondrial membrane.

Section 22.8 Complete Oxidation of Glucose

One molecule of cytoplasmic NADH in the muscles generates (a) _____ molecules of ATP, while (b) _____ molecules of ATP come from each molecule of mitochondrial NADH. The complete aerobic catabolism of 1 mol of glucose in the liver produces (c) _____ molecules of ATP. The majority of ATP molecules are formed as a result of oxidative (d) _____.

Section 22.9 Glycogen Metabolism

The synthesis of glycogen from glucose is called (a) _____. Energy for the synthesis of glycogen is provided by the high-energy compound (b) _____. The breakdown of glycogen to glucose is called (c) _____. The enzyme glucose 6-phosphatase which is necessary for the conversion of glycogen to glucose is found primarily in the (d) _____.

Section 22.10 Gluconeogenesis

Gluconeogenesis is the synthesis of (a) _____ from noncarbohydrate molecules. The majority of gluconeogenesis takes place in the (b) _____. A key intermediate in the conversion of lactate to glucose is (c) _____. The cyclic process in which glucose is converted to lactate and lactate is reconverted to glucose is called the (d) _____ cycle.

Section 22.11 Hormonal Control of Carbohydrate Metabolism

Insulin enhances the absorption of (a) _____ from blood into cells. The hormone (b) _____ works in opposition to insulin by raising blood glucose levels. Epinephrine stimulates (c) _____ breakdown in the muscles.

SOLUTIONS TO EXERCISES ANSWERED IN THE TEXT

22.2   glucose, galactose, and fructose

22.4   Blood sugar level is the concentration of glucose in the blood. The normal fasting level is the blood glucose concentration after a fast of 8-12 hours.

22.6   b) a blood sugar level above the normal fasting level
       d) an elevated blood sugar level resulting in the excretion of glucose in the urine

22.7   Severe hypoglycemia can cause convulsions and shock.

22.9   to start the breakdown of carbohydrates for energy production

22.11 cytoplasm

22.13 NAD^+ in step 6

22.16 High concentrations of glucose 6-phosphate inhibit the first step of glycolysis.

22.18 Aerobic denotes the presence of oxygen, whereas anaerobic is without oxygen.

22.20 Lactate formation from pyruvate regenerates a supply of NAD^+, which is necessary for glycolysis to continue.

22.21 b) $C_6H_{12}O_6 + 2ADP + 2P_i + 2NAD^+ \rightarrow$

$$H_2O + 2CH_3\overset{\overset{\displaystyle O}{\|}}{C}-COO^- + 2ATP + 2NADH + 2H^+$$

22.22 The two ATPs produced are sufficient to maintain life.

22.24 The process starts with the reaction of oxaloacetate, and the final step of the pathway regenerates the oxaloacetate.

22.25 b) CO_2
c) NADH and $FADH_2$

22.26 The reduced coenzyme products of the citric acid cycle are the fuel molecules for the electron transport chain.

22.28 b) one
d) two

22.30
$$\begin{array}{c}COO^- \\ | \\ CH_2 \\ | \\ CH_2 \\ | \\ COO^-\end{array} + FAD \rightarrow \begin{array}{c}H\quad COO^- \\ \diagdown \diagup \\ C \\ \| \\ C \\ \diagup \diagdown \\ ^-OOC \quad H\end{array} + FADH_2$$

succinate   fumarate

22.32 b) When supplies of NADH and ATP are abundant, the cell has sufficient energy for its needs, and so the operation of the citric acid cycle is inhibited.

22.34 NADH and $FADH_2$

22.36 cytochromes

22.38 NADH and $FADH_2$ are oxidized to NAD^+ and FAD, respectively. ADP is phosphorylated to ATP.

22.40 Two molecules of ATP are formed during a sequence of reactions.

22.42 One turn of the citric acid cycle produces 1 $FADH_2$, 3NADH + $3H^+$, and 1 GTP. Each $FADH_2$ gives rise to 2 molecules of ATP in the electron transport chain, each NADH + H^+ gives 3 molecules of ATP, and GTP is considered equivalent to ATP. Thus, one turn of the citric acid cycle coupled with the electron transport chain produces 12 molecules of ATP. Six acetyl CoA molecules would fuel six turns of the citric acid cycle and produce 6 x 12 or 72 ATP molecules.

22.44 The synthesis of ATP from ADP requires 7.3 kcal/mol of energy. 686 kcal could potentially fuel the synthesis of 686 kcal/7.3 kcal or 94 moles of ATP.

22.46 F$_1$-ATPase provides a channel for H$^+$ ions to flow back across the inner membrane.

22.48 a) Glycolysis gives 2 mol ATP directly and 2 moles of NADH which must go to the electron transport chain.

b) The citric acid cycle produces 1 mole of GTP which is equivalent to one mole of ATP. The FADH$_2$ and NADH produced go on to the electron transport chain. Because 1 mole of glucose gives 2 moles of acetyl CoA, the citric acid cycle goes around twice to give 2 moles of GTP or 2 moles ATP.

c) Thus, glycolysis gives 2ATP, the citric acid cycle gives 2 ATP, and the electron transport chain with oxidative phosphorylation provides the remainder or 34 moles of ATP.

22.49 a) Glycolysis produces 2 moles of ATP. Thus, 2 mole x 7.3 kcal/mole or 14.6 kcal or energy is conserved.

$$\frac{14.6\ kcal}{686\ kcal} \times 100 = 2.13\%\ conserved$$

22.51 UTP

22.53 The enzyme glucose 6-phosphatase, which is necessary to produce free glucose, is found in the liver but not in muscle cells.

22.55 The breakdown of glycogen involves the conversion to glucose 1-phosphate, which forms glucose 6-phosphate, followed by the formation of glucose.

22.57 Lactate (from anaerobic oxidation of glucose), glycerol (from fats), and certain amino acids (from proteins)

22.59 b) Insulin promotes glycogen formation, and glucagon promotes the breakdown of glycogen.

22.61 Lactose intolerance is caused by a deficiency of lactase and is characterized by bloating, stomach cramps, gas, or diarrhea.

22.64 Strenuous exercise results in a slight decrease in blood pH, which stimulates the body to breathe heavily bringing in more oxygen.

22.66 In juvenile-onset diabetes, which usually appears in children before the age of 10, practically no insulin is produced. Maturity-onset diabetes results from a gradual decrease in insulin production.

22.68 Hyperglycemia, glucosuria, and a lowered blood pH may develop; if left unchecked, this can eventually lead to blindness, coma, and death.

SELF-TEST QUESTIONS

Multiple Choice

1. The digestion of carbohydrates produces glucose, fructose and
   a) starch
   b) amylose
   c) lactose
   d) galactose

2. The central compound in carbohydrate metabolism is
   a) glycogen
   b) glucose
   c) fructose
   d) galactose

3. The glycolysis pathway is located within the _____ of cells.
   a) cytoplasm
   b) nucleus
   c) mitochondria
   d) ribosomes

4. The net production of ATP in glycolysis from one molecule of glucose is _____ molecules.
   a) 0
   b) 1
   c) 2
   d) 4

5. The net production of NADH in glycolysis from one molecule of glucose is _____ molecules.
   a) 0
   b) 1
   c) 2
   d) 4

6. Enzymes regulate the glycolysis pathway at _____ control points.
   a) 2
   b) 3
   c) 4
   d) 5

7. The muscle pain which follows prolonged and vigorous contraction of skeletal muscles is the result of the accumulation of
   a) lactate
   b) citrate
   c) NADH
   d) pyruvate

8. How many molecules of pyruvate are produced from the glycolysis of one molecule of glucose?
   a) 0
   b) 1
   c) 2
   d) 3

9. How many molecules of ATP can be formed in the muscles from every molecule of cytoplasmic NADH produced during glycolysis?
   a) 0   b) 1   c) 2   d) 3

10. The complete oxidation of glucose in the liver results in the formation of _____ molecules of ATP.
    a) 18   b) 32   c) 34   d) 38

11. Which of the following exerts an effect on blood sugar levels opposite to that of insulin?
    a) cholesterol
    b) glucagon
    c) vasopressin
    d) aldosterone

12. What disease is commonly associated with glucosuria?
    a) diabetes mellitus
    b) hepatitis
    c) hypoglycemia
    d) hyperinsulinism

13. The first step of the citric acid cycle involves the reaction of oxaloacetate with ___ to form citrate.
    a) coenzyme A
    b) acetyl CoA
    c) CO_2
    d) malate

14. The high-energy compound formed in the citric acid cycle is
    a) CTP   b) ATP   c) UTP   d) GTP

15. One turn of the citric acid cycle produces _____ molecules of NADH.
    a) 1   b) 2   c) 3   d) 4

16. The citric acid cycle is inhibited by
    a) ATP   b) CO_2 c) NAD^+ d) FAD

17. One product of the electron transport chain is
    a) O_2 b) H_2O c) CO_2 d) NADH

18. How many sites in the electron transport chain can support the synthesis of ATP?
    a) 1   b) 2   c) 3   d) 4

Matching

Match each description on the left to the conditions or diseases on the right.

19. a high glucose level in the blood
20. a low glucose level in the blood
21. a high glucose level in the urine

a) glucosuria
b) hypoglycemia
c) hyperglycemia
d) galactosemia

Match each description on the left to the products on the right.

22. a product of fermentation
23. produced by oxidation under aerobic conditions
24. produced by glycolysis under anaerobic conditions

a) acetaldehyde
b) lactate
c) acetyl CoA
d) ethanol

Match each characteristic on the left to the cycle or process on the right.

25. glucose is converted to glycogen
26. synthesis of glucose from noncarbohydrate sources
27. breakdown of glycogen to glucose

a) glycogenesis
b) gluconeogenesis
c) glycogenolysis
d) glycolysis

True-False

28. Fructose and galactose are not metabolized by humans.

29. The blood sugar level in a hypoglycemic individual is higher than the normal fasting level.

30. Some ATP is formed during glycolysis.

31. The glycolysis pathway is inhibited by ATP.

32. Part of the Cori cycle involves the conversion of pyruvate to glucose.

33. About 90% of gluconeogenesis takes place in the liver.

34. Oxidative phosphorylation is a process coupled with the electron transport chain.

35. The chemiosmotic hypothesis pertains to the flow of cytoplasmic NADH across the mitochondrial membrane.

36. The net production of ATP from one mole of glucose in brain cells is greater than that in liver cells.

ANSWERS TO PROGRAMMED REVIEW

22.1 a) energy   b) monosaccharides   c) glucose

22.2 a) level   b) hypoglycemia   c) hyperglycemia
      d) renal threshold

22.3 a) pyruvate   b) cytoplasm   c) enzymes   d) feedback

22.4 a) acetyl CoA   b) lactate   c) ethanol   d) NAD^+

22.5 a) acetyl CoA   b) CO_2 c) three   d) one

22.6 a) mitochondrion b) H_2O c) cytochromes   d) NADH

22.7 a) three   b) ATP   c) chemiosmotic   d) proton

22.8 a) two   b) three   c) 38   d) phosphorylation

22.9 a) glycogenesis   b) UTP   c) glycogenolysis   d) liver

22.10 a) glucose   b) liver   c) pyruvate   d) Cori

22.11 a) glucose   b) glucagon   c) glycogen

ANSWERS TO SELF-TEST QUESTIONS

1.	d	13.	b	25.	a
2.	b	14.	d	26.	b
3.	a	15.	c	27.	c
4.	c	16.	a	28.	F
5.	c	17.	b	29.	F
6.	b	18.	c	30.	T
7.	a	19.	c	31.	T
8.	c	20.	b	32.	T
9.	c	21.	a	33.	T
10.	d	22.	b	34.	T
11.	b	23.	c	35.	F
12.	a	24.	b	36.	F

CHAPTER 23

Lipid and Amino Acid Metabolism

PROGRAMMED REVIEW

Section 23.1 Blood Lipids

Lipoprotein aggregates found in the lymph and blood are called (a) _____. Lipoproteins with the greatest density are called (b) _____ lipoproteins. The protein component is greatest in the (c) _____ lipoproteins. The cholesterol content of lipoproteins is useful in assessing the risk of (d) _____.

Section 23.2 Fat Mobilization

Triglycerides are stored in (a) _____ tissue. Fat mobilization involves the (b) _____ of triglycerides followed by the entry of fatty acids and (c) _____ into the bloodstream.

Section 23.3 Glycerol Metabolism

Glycerol is converted to (a) _____ phosphate, one of the intermediates of the (b) _____ pathway. Glycerol can be converted into pyruvate for energy production or to (c) _____ through gluconeogenesis.

Section 23.4 Oxidation of Fatty Acids

A fatty acid is activated by a reaction with (a) _____ in the presence of ATP. β-oxidation of a fatty acid produces molecules of (b) _____. Every run through the β-oxidation process produces the reduced coenzymes (c) _____ and (d) _____.

Section 23.5 The Energy from Fatty Acids

Activation of a fatty acid requires the equivalent of hydrolysis of (a) _____ molecules of ATP to ADP. A ten-carbon fatty acid requires (b) _____ trips through the fatty acid spiral and produces (c) _____ molecules of acetyl CoA and (d) _____ molecules each of FADH$_2$ and NADH.

Section 23.6 Ketone Bodies

Ketone bodies are synthesized from (a) _____. The ketone body which can sometimes be detected on the breath of a diabetic is (b) _____. An abnormally low blood pH due to the presence of ketone bodies is called (c) _____. Ketonuria is the presence of ketone bodies in the (d) _____.

Section 23.7 Fatty Acid Synthesis

Biosynthesis of fatty acids occurs within the (a) _____ of the cell. Intermediates during fatty acid synthesis are attached to an acyl carrier (b) _____. The (c) _____ is the most important organ involved in fatty acid synthesis.

Section 23.8 Amino Acid Metabolism

In terms of amount used, the most important function of amino acids is to provide building blocks for the synthesis of (a) _____ in the body. The dynamic process in which body proteins are continuously hydrolyzed and resynthesized is called protein (b) _____. Unlike carbohydrates and fatty acids, amino acids in excess of immediate body requirements cannot be (c) _____ for later use. The turnover rate for proteins is usually expressed as a (d) _____-life.

Section 23.9 Catabolism of Amino Acids: Fate of the Nitrogen Atoms

Enzymes which catalyze the transfer of amino groups are called (a) _____. The term oxidative (b) _____ is applied to an oxidation process resulting in the removal of an amino group. The organ where the urea cycle is located is the (c) _____. The fuel for the urea cycle is (d) _____ phosphate.

Section 23.10 Catabolism of the Carbon Skeleton of Amino Acids

(a) _____ amino acids are those whose carbon skeletons can be converted to intermediates used in the synthesis of glucose. (b) _____ amino acids are those whose carbon skeletons can be converted to acetyl CoA or acetoacetyl CoA. All 20 amino acids can be degraded into pyruvate, acetyl CoA, acetoacetyl CoA or intermediates of the (c) _____ cycle.

Section 23.11 Amino Acid Biosynthesis

Amino acids that can be synthesized in the amounts needed by the body are called (a) _____ amino acids. Amino acids which cannot be made in large enough amounts to meet bodily needs must be included in our diet and are called (b) _____ amino acids. The key starting materials for the synthesis of nine amino acids are intermediates of the (c) _____ pathway and the citric acid cycle. Three amino acids are synthesized from α-keto acids via reactions catalyzed by (d) _____.

SOLUTIONS TO EXERCISES ANSWERED IN THE TEXT

23.2 Glycerol, fatty acids, and monoglycerides

23.4 Lipids are transported as lipoprotein complexes.

23.6 chylomicrons, very low-density lipoproteins, low-density lipoproteins, high-density lipoproteins

23.8 The hormone epinephrine initiates fat mobilization.

23.10 resting muscle and liver cells

23.12 Glycerol can be converted to pyruvate and contribute to cellular energy production or be converted to glucose.

23.14 A fatty acid is first converted to a fatty acyl CoA, a high-energy compound. This occurs in the cytoplasm of the cell.

23.16 FAD and NAD$^+$

23.18
$$CH_3(CH_2)_{14}-\overset{O}{\underset{\|}{C}}-OH + HS-CoA \xrightarrow{ATP \rightarrow AMP + 2P_i} CH_3(CH_2)_{14}-\overset{O}{\underset{\|}{C}}-SCoA + H_2O$$

23.20 The word *cycle* implies that the product is identical to the starting material, which is not the case. The fatty acyl product of a trip through the sequence has two fewer carbon atoms than the starting material.

23.22 *Step 1:* a dehydrogenation to produce a carbon-carbon double bond
Step 2: hydration to produce an alcohol
Step 3: oxidation to produce a ketone
Step 4: cleavage to produce acetyl CoA and a new fatty acyl CoA

23.23 b) $CH_3CH_2CH_2CH=CH-\overset{O}{\underset{\|}{C}}-SCoA + H_2O \rightarrow CH_3CH_2CH_2\overset{OH}{\underset{|}{CH}}CH_2-\overset{O}{\underset{\|}{C}}-SCoA$

d) $CH_3CH_2CH_2\overset{O}{\underset{\|}{C}}CH_2-\overset{O}{\underset{\|}{C}}-SCoA + CoA-SH \rightarrow CH_3CH_2CH_2\overset{O}{\underset{\|}{C}}-SCoA + CH_3-\overset{O}{\underset{\|}{C}}-SCoA$

23.24 Two high-energy bonds of ATP (the equivalent of two ATP molecules) are used in the activation of a fatty acid.

23.26 A ten-carbon fatty acid would produce 5 molecules of acetyl CoA. It would require 4 trips through the fatty acid spiral, producing 4 $FADH_2$ and 4 NADH molecules. Each $FADH_2$ gives 2 ATP molecules in the electron transport chain; each NADH gives 3 ATP molecules. Every acetyl CoA gives 12 ATP molecules. Adding each of these together gives

5 acetyl CoA	x	12 ATP	=	60 ATP
4 $FADH_2$	x	2 ATP	=	8 ATP
4 NADH	x	3 ATP	=	12 ATP
				80 ATP

Two ATP molecules are subtracted for the initial activation step, giving a net 78 ATP.

23.28 Ketone bodies are three compounds formed from acetyl CoA: acetoacetate, β-hydroxybutyrate, and acetone.

23.29 b) In patients with diabetes mellitus, cells are not able to sufficiently metabolize glucose. As the body switches to fat metabolism, greater amounts of acetyl CoA are produced. Increased acetyl CoA results in greater production of ketone bodies.

23.31 a) Ketonemia refers to high amounts of ketone bodies in the blood.
b) Ketonuria refers to the presence of ketone bodies in the urine.
c) Acetone breath is the condition where the concentration of acetone in the blood reaches levels that cause it to be expelled through the lungs.
d) Ketosis is the condition when ketonemia, ketonuria, and acetone breath exist simultaneously.

23.33 cytoplasm

23.35 ATP and NADPH

Lipid and Amino Acid Metabolism 295

23.38 The amino acid pool is a supply of amino acids located within the blood and cellular spaces.

23.40 Protein turnover refers to the degradation and resynthesis of body proteins.

23.42 purines and pyrimidines, heme, choline, ethanolamine

23.44 b) deamination
d) transamination and deamination

23.47 $\text{C}_6\text{H}_5-\text{CH}_2-\overset{\overset{\displaystyle O}{\|}}{\text{C}}-\text{COO}^- + \text{NADH} + \text{H}^+ + \text{NH}_4^+$

23.49 a) $\underset{\text{alanine}}{\text{CH}_3-\overset{\overset{\displaystyle NH_3^+}{|}}{\text{CH}}-\text{COO}^-} + {}^-\text{OOC}-\text{CH}_2\text{CH}_2-\overset{\overset{\displaystyle O}{\|}}{\text{C}}-\text{COO}^- \xrightarrow{\text{transaminase}}$

$\text{CH}_3-\overset{\overset{\displaystyle O}{\|}}{\text{C}}-\text{COO}^- + \underset{\text{glutamate}}{{}^-\text{OOC}-\text{CH}_2\text{CH}_2-\overset{\overset{\displaystyle NH_3^+}{|}}{\text{CH}}-\text{COO}^-}$

b) $\underset{\text{glutamate}}{{}^-\text{OOC}-\text{CH}_2\text{CH}_2-\overset{\overset{\displaystyle NH_3^+}{|}}{\text{CH}}-\text{COO}^-} + \text{NAD}^+ + \text{H}_2\text{O} \xrightarrow{\text{glutamate dehydrogenase}}$

$\text{NH}_4^+ + \text{NADH} + \text{H}^+ + {}^-\text{OOC}-\text{CH}_2\text{CH}_2-\overset{\overset{\displaystyle O}{\|}}{\text{C}}-\text{COO}^-$

23.51 The carbon atom is from CO_2. One nitrogen atom is from NH_4^+ (from the deamination of glutamate), and one is from aspartate.

23.53 Three ATP molecules are used, producing two ADP and one AMP molecules.

23.55 Glucogenic amino acids can be degraded into pyruvate or intermediates of the citric acid cycle and used for the synthesis of glucose. Ketogenic amino acids cannot be used for glucose production and are converted into acetyl CoA or acetoacetyl CoA.

23.57 acetyl CoA, α-ketoglutarate, succinyl CoA, fumarate, oxaloacetate

296 CHAPTER 23

23.59 b) ketogenic

23.60 Amino acids that can be synthesized by the body are called nonessential. Those amino acids that must be obtained from the diet because the body cannot produce them are called essential amino acids.

23.62 the glycolysis pathway and the citric acid cycle

23.64 through exercise, weight loss, or drug therapy

23.66 Increasing the length of the exercise period prolongs the postexercise higher rate of using energy.

23.68 hyperammonemia

23.70 The diet of an infant with PKU is controlled to eliminate phenylalanine and to supplement with tyrosine.

SELF-TEST QUESTIONS

Multiple Choice

1. What reaction occurs during the digestion of triglycerides?
   a) hydrolysis
   b) hydrogenation
   c) hydration
   d) oxidation

2. The products of triglyceride digestion are fatty acids, some monoglycerides, and
   a) triglycerides
   b) diglycerides
   c) glucose
   d) glycerol

3. Lipoproteins which contain the greatest amount of protein are the
   a) chylomicrons
   b) very-low density lipoproteins
   c) low-density lipoproteins
   d) high-density lipoproteins

4. What reaction occurs during fat mobilization?
   a) reduction
   b) hydrolysis
   c) oxidation
   d) hydration

5. One of the hormones involved in fat mobilization is
   a) aldosterone
   b) vasopressin
   c) insulin
   d) epinephrine

6. The entry point for glycerol into the glycolysis pathway is
   a) dihydroxyacetone phosphate
   b) acetyl CoA
   c) pyruvate
   d) fructose 1,6-diphosphate

7. In order for fatty acids to enter the mitochondria for degradation, they must first be converted to
   a) acetyl CoA
   b) fatty acyl CoA
   c) pyruvate
   d) malonate

8. Which of the following processes takes place during the fatty acid spiral?
   a) addition of water to a double bond
   b) oxidation of an OH group to a ketone
   c) addition of H to a double bond
   d) more than one response is correct

9. During one run through the fatty acid spiral, which bond of the following fatty acid would be broken?

$$CH_3(CH_2)_4-CH_2-CH_2-CH_2-CH_2-\overset{O}{\overset{\|}{C}}-OH$$
$$\underset{d}{\uparrow}\underset{c}{\uparrow}\underset{b}{\uparrow}\underset{a}{\uparrow}$$

   a) bond a
   b) bond b
   c) bond c
   d) bond d

10. Which of the following is a product of the fatty acid spiral?
    a) pyruvate
    b) acetyl CoA
    c) $CO_2 + H_2O$
    d) more than one response is correct

11. How many runs through the fatty acid spiral would be required to completely break down one molecule of a 12-carbon fatty acid?
    a) a
    b) 5
    c) 6
    d) 12

12. How many FADH$_2$ molecules are produced from one turn of the fatty acid spiral?
    a) 1
    b) 2
    c) 3
    d) 4

13. How many ATP molecules ultimately result from one turn of the fatty acid spiral?
    a) 8
    b) 12
    c) 14
    d) 17

298 CHAPTER 23

14. The concentration of ketone bodies builds up when increased amounts of _____ are oxidized.
    a) amino acids
    b) fatty acids
    c) glucose
    d) glycerol

15. Which of the following types of proteins tends to have the shortest half-life in the body?
    a) enzymes
    b) plasma proteins
    c) connective tissue proteins
    d) muscle proteins

16. The primary function of the urea cycle in the body is to
    a) produce ATP from ADP
    b) convert amino acids into keto acids
    c) convert keto acids into amino acids
    d) convert ammonium ions into urea

17. Which of the following substances is a product of a transamination reaction?
    a) amino acid
    b) ammonia
    c) keto acid
    d) more than one response is correct

18. The nitrogen of amino acids appears in the urine of mammals primarily as
    a) uric acid
    b) urea
    c) ammonia
    d) N_2

19. Which of the following substances enters into the urea cycle?
    a) malonyl CoA
    b) acetyl CoA
    c) phosphocreatine
    d) carbamoyl phosphate

20. The carbon atom of urea is derived from
    a) CO_2
    b) acetyl CoA
    c) pyruvate
    d) aspartate

21. Deamination of an amino acid produces
    a) ammonium ions
    b) an α-keto acid
    c) CO_2
    d) more than one response is correct

22. The carbon skeletons of amino acids are ultimately catabolized through
    a) glycolysis
    b) glycogenolysis
    c) the citric acid cycle
    d) the urea cycle

23. A number of amino acids can be synthesized in the body from intermediates of
    a) the citric acid cycle
    b) the urea cycle
    c) oxidative phosphorylation
    d) the fatty acid spiral

Lipid and Amino Acid Metabolism

True-False

24. Fat has a caloric value more than twice that of glycogen and starch.

25. Upon complete oxidation to CO_2 and H_2O, fatty acids produce more net energy than a carbohydrate containing the same number of carbon atoms.

26. The glycerol resulting from triglyceride hydrolysis can be converted to pyruvate.

27. The glycerol resulting from triglyceride hydrolysis can be converted to glucose.

28. Starvation can lead to ketosis.

29. Ketosis results when too little acetyl coenzyme A is produced for the needs of the body.

30. Amino acids in excess of immediate body requirements are stored for later use.

31. Amino acids can be catabolized for energy production.

32. Urea formation is an energy-yielding process.

33. The human body excretes small amounts of nitrogen as ammonium ions.

34. The carbon atoms of some amino acids can be used in fatty acid synthesis.

35. The carbon atoms of some amino acids can be used in glucose synthesis.

36. Biosynthesis of fatty acids occurs in the mitochondria of the cell.

37. The liver is the most important organ in fatty acid synthesis.

ANSWERS TO PROGRAMMED REVIEW

23.1   a) chylomicrons   b) high-density   c) high-density
       d) heart attack

23.2   a) adipose   b) hydrolysis   c) glycerol

300   CHAPTER 23

23.3   a) dihydroxyacetone   b) glycolysis   c) glucose

23.4   a) coenzyme A   b) acetyl CoA   c) FADH$_2$ d) NADH

23.5   a) two   b) four   c) five   d) four

23.6   a) acetyl CoA   b) acetone   c) ketoacidosis   d) urine

23.7   a) cytoplasm   b) protein   c) liver

23.8   a) protein   b) turnover   c) stored   d) half

23.9   a) transaminases   b) deamination   c) liver   d) carbamoyl

23.10 a) glucogenic   b) ketogenic   c) citric acid

23.11 a) nonessential   b) essential   c) glycolysis   d) transaminases

ANSWERS TO SELF-TEST QUESTIONS

1.	a	14.	b	26.	T
2.	d	15.	a	27.	T
3.	d	16.	d	28.	T
4.	b	17.	d	29.	F
5.	d	18.	b	30.	F
6.	a	19.	d	31.	T
7.	b	20.	a	32.	F
8.	d	21.	d	33.	T
9.	b	22.	c	34.	T
10.	b	23.	a	35.	T
11.	b	24.	T	36.	F
12.	a	25.	T	37.	T
13.	d				

CHAPTER 24

Body Fluids

PROGRAMMED REVIEW

Section 24.1 Comparison of Body Fluids

The majority of body fluids are located inside the cells and are called (a) _____ fluid. All body fluids not located inside the cells are collectively known as (b) _____ fluids. Chemically, the two extracellular fluids, plasma and (c) _____ fluid, are nearly identical. The principal cation of plasma is (d) _____.

Section 24.2 Oxygen and Carbon Dioxide Transport

The oxygenated form of hemoglobin is called (a) _____. Hemoglobin combined with CO_2 is known as (b) _____. The majority of CO_2 is carried from body tissues to the lungs in the form of (c) _____ ions. The movement of chloride ions to maintain electrical neutrality within red blood cells is called the chloride (d) _____.

Section 24.3 Chemical Transport to the Cells

Fluid flow through capillary walls is governed by blood pressure and by (a) _____ pressure. Of these two factors, (b) _____ pressure is greater at the arterial end of a capillary and there is a tendency for a net flow of fluid to occur (c) _____ the capillary.

Section 24.4 Constituents of Urine

The presence of large amounts of bile pigments in urine may be indicative of (a) _____. Ketonuria is a term used to describe the presence of (b) _____ within urine. The cation present in greatest amounts in urine is (c) _____. The anion most prevalent in urine is (d) _____.

Section 24.5 Fluid and Electrolyte Balance

Water intake is regulated by the (a) _____ mechanism. Water leaves the body through the intestines, kidneys, skin and (b) _____. (c) _____ is known as the antidiuretic hormone. The hormone (d) _____ stimulates the reabsorption of Na^+ ions.

301

Section 24.6 Acid-Base Balance

Blood pH is normally within the range of 7.35 to (a) _____. An increase in blood pH is called (b) _____. An abnormally low blood pH is called (c) _____. A constant blood pH is maintained by the interactive operation of three systems: buffer, (d) _____, and urinary.

Section 24.7 Buffer Control of pH

Three major buffer systems of the blood are the bicarbonate buffer, the (a) _____ buffer, and the (b) _____ proteins. The most important of these is the bicarbonate buffer system, consisting of a mixture of HCO_3^- and (c) _____.

Section 24.8 Respiratory Control of pH

The water and carbon dioxide which are exhaled are formed from (a) _____ acid. An increased rate of breathing called (b) _____ is caused by a (c) _____ blood pH. Slow, shallow breathing is called (d) _____.

Section 24.9 Urinary Control of pH

The excretion of H^+ within the urine is accompanied by the conversion of CO_2 to (a) _____ within the blood. For every H^+ ion entering the urine, a (b) _____ ion passes into the tubule cells. The enzyme which catalyzes the combining of CO_2 and H_2O is (c) _____.

Section 24.10 Acidosis and Alkalosis

When blood pH is normal and balanced, it contains 20 parts of bicarbonate ions to 1 part of (a) _____ acid. A condition of (b) _____ alkalosis is caused by hyperventilation. (c) _____ acidosis is a condition of acidosis resulting from causes other than hypoventilation. Excessive intake of baking soda may give rise to metabolic (d) _____.

SOLUTIONS TO EXERCISES ANSWERED IN THE TEXT

24.2   plasma and interstitial fluid

24.4   b) intracellular
        d) intracellular
        f) intracellular

Body Fluids 303

24.6 nucleic acids and metabolic intermediates

24.7 b) oxyhemoglobin

24.9 b) 98%
d) 5%

24.10 Red blood cells are the site for all of the reactions both within the lungs and at the tissue cells.

24.12 $HHb + O_2 \rightleftharpoons HbO_2^- + H^+$
b) right
d) left

24.15 a) osmotic pressure
b) blood pressure

24.18 if the concentration were abnormally high

24.20 the metabolism of food molecules

24.22 variations in urine output

24.26 blood buffers, urinary control, and respiratory control

24.28 $HPO_4^{2-} + H^+ \rightarrow H_2PO_4^-$
$H_2PO_4^- + OH^- \rightarrow H_2O + HPO_4^{2-}$

24.30 Exhaled H_2O and CO_2 are derived from H_2CO_3. The more CO_2 and H_2O that are exhaled, the more carbonic acid is removed from the blood, thus elevating the blood pH.

24.32 A blood pH lower than normal has caused an increased rate of breathing to exhale more H_2O and CO_2, decreasing the amount of carbonic acid in the blood.

24.34 CO_2 diffuses into the distal tubule cells to form carbonic acid, which then ionizes to H^+ and HCO_3^-. Thus, CO_2 concentration is lowered in the blood.

24.36 As H^+ enters the urine, Na^+ is reabsorbed into the blood from the urine.

24.38 Respiratory acidosis is caused by hypoventilation. Respiratory alkalosis is caused by hyperventilation.

304 CHAPTER 24

24.40 Vomiting results in a loss of acid from the body, which can lead to a rise in blood pH (alkalosis).

24.42 Osteoporosis is characterized by a decrease in bone calcium and brittle bones.

24.44 Consume foods rich in calcium and vitamin D and get plenty of exercise.

24.46 Some sports drinks contain carbohydrates for energy, which might not be of value if the goal is to lose weight.

24.48 Heat stroke is prevented by drinking plenty of fluids, resting in the shade, and wearing lightweight clothing.

SELF-TEST QUESTIONS

Multiple Choice

1. Which of the following fluids contain similar concentrations of Na^+ and Cl^- ions?
   a) blood plasma and interstitial fluid
   b) blood plasma and intracellular fluid
   c) interstitial fluid and intracellular fluid
   d) all three have about the same concentrations

2. A body fluid is analyzed and found to contain a low concentration of Na^+ and a high concentration of K^+. The fluid is most likely
   a) blood plasma
   b) intracellular fluid
   c) interstitial fluid
   d) all three have about the same concentrations

3. Which of the following reactions takes place in red blood cells at the lungs?
   a) $H_2CO_3 \rightarrow H_2O + CO_2$
   b) $CO_2 + H_2O \rightarrow H_2CO_3$
   c) $H^+ + HbO_2^- \rightarrow HHb + O_2$
   d) $H_2CO_3 \rightarrow H^+ + HCO_3^-$

4. Most oxygen is carried to various parts of the body, via the bloodstream, in the form of
   a) a dissolved gas
   b) oxyhemoglobin
   c) bicarbonate ion
   d) carbon dioxide

5. During respiration reactions at the lungs and cells, the function of the chloride shift is to
   a) eliminate toxic chlorine from the body
   b) maintain pH balance in red blood cells
   c) maintain charge balance in red blood cells
   d) activate the carbonic anhydrase enzyme in red blood cells

6. Which of the following would be considered to be an abnormal constituent in urine?
   a) ammonium ion
   b) creatinine
   c) protein
   d) bicarbonate salts

7. The vasopressin mechanism and the aldosterone mechanism both tend to regulate
   a) the fluid and electrolyte levels in the body
   b) the rate of glucose oxidation in the body
   c) the rate of hemoglobin production in the body
   d) the CO_2 levels in the cells

8. Which of the following maintains a constant pH for the blood?
   a) respiration reactions associated with breathing
   b) kidney activity
   c) formation and excretion of perspiration
   d) more than one response is correct

9. Which buffer system is regulated in part by the kidneys and by the respiratory system?
   a) bicarbonate
   b) phosphate
   c) protein
   d) ammonium

10. The three major buffer systems of the blood are the plasma proteins, the bicarbonate buffer, and the
    a) succinate buffer
    b) ammonium buffer
    c) lactate buffer
    d) phosphate buffer

11. Exhaling CO_2 and H_2O
    a) raises blood pH
    b) lowers blood pH
    c) increases the blood concentration of H_2CO_3
    d) has no effect on H_2CO_3 concentration

12. Blood pH normally remains in the range
    a) 7.15 - 7.25
    b) 7.25 - 7.35
    c) 7.35 - 7.45
    d) 7.45 - 7.55

Matching

For each cause of blood acid-base imbalance listed on the left, select the resulting condition from the responses.

13.   prolonged diarrhea

14.   excessive intake of baking soda

15.   hypoventilation

16.   hyperventilation

a) respiratory acidosis
b) respiratory alkalosis
c) metabolic acidosis
d) metabolic alkalosis

For each condition listed on the left, identify the resulting abnormal urine constituent on the right.

17.   hepatitis

18.   starvation

19.   diabetes mellitus

a) ketone bodies
b) glucose (in large amounts)
c) bile pigments
d) more than one constituent results

Select a correct name for each formula on the left.

20. HbO_2^-

21.   HHb

22. $HHbCO_2$

a) carbaminohemoglobin
b) oxyhemoglobin
c) deoxyhemoglobin
d) carboxyhemoglobin

True-False

23. Plasma is classified as an intracellular fluid.

24. Osmotic pressure differences between plasma and interstitial fluid always tend to move fluid into the blood.

25. Osmotic pressure exceeds heart pressure at the venous end of the circulatory system.

26. An increased rate of breathing tends to lower blood pH.

27. The excretion of H^+ ions decreases urine pH.

28. The pH of blood is increased by the excretion of H⁺ in urine.

29. Vomiting may give rise to metabolic acidosis.

30. The concentration of protein in plasma is much higher than the protein concentration in the interstitial fluid.

31. The body is 82-90% water.

32. Aldosterone is also called the antidiuretic hormone (ADH).

33. Death can result if blood pH falls below 6.8.

34. H_2CO_3 is a moderately strong acid.

35. Urine is buffered by the phosphate buffer.

ANSWERS TO PROGRAMMED REVIEW

24.1	a) intracellular	b) extracellular	c) interstitial	d) Na⁺
24.2	a) oxyhemoglobin d) shift	b) carbaminohemoglobin	c) HCO_3^-	
24.3	a) osmotic	b) blood	c) from	
24.4	a) jaundice	b) ketone bodies	c) Na⁺	d) Cl⁻
24.5	a) thirst	b) lungs	c) vasopressin	d) aldosterone
24.6	a) 7.45	b) alkalosis	c) acidosis	d) respiratory
24.7	a) phosphate	b) plasma	c) H_2CO_3	
24.8	a) carbonic	b) hyperventilation	c) low	d) hypoventillation
24.9	a) HCO_3^-	b) Na⁺	c) carbonic anhydrase	
24.10	a) carbonic	b) respiratory	c) metabolic	d) alkalosis

ANSWERS TO SELF-TEST QUESTIONS

1.	a	13.	c	25.	T
2.	b	14.	d	26.	F
3.	a	15.	a	27.	T
4.	b	16.	b	28.	T
5.	c	17.	c	29.	F
6.	c	18.	a	30.	T
7.	a	19.	d	31.	F
8.	d	20.	b	32.	F
9.	a	21.	c	33.	T
10.	d	22.	a	34.	F
11.	a	23.	F	35.	T
12.	c	24.	T		